T0130910

MASSEY FERGUSON 100 SERIES In Detail

MASSEY FERGUSON
100 SERIES
In Detail

BY MICHAEL THORNE

Photography by Andrew Morland

Herridge & Sons

Published 2017 by
Herridge & Sons Ltd
Lower Forda, Shebbear
Beaworthy,
Devon EX21 5SY

ISBN 978-1-906133-76-4
Printed in China

Contents

An Introduction to the MF100 Series of Tractors

The decision to proceed with the development of the MF100 Series tractors was made at a corporate co-ordinating committee meeting in Detroit in late 1962. It was decided that the company would not continue to buy in larger tractors from outside manufacturers like Oliver and Minneapolis Moline, but instead would develop in-house its own larger-horsepower range of tractors incorporating the famous Ferguson System hydraulics. In addition to Draft and Position control, on all tractors across the range the system would incorporate an innovation in its design and function which would enable the weight transfer principle to be applied to trailed implements and trailer; this became known as Pressure Control. This function was not available on the French built MF130 but was optional on earlier models. It is interesting to note that the Swedish MF dealer, A.B. Nykopings Automobilfabrik, had developed a similar but more elementary device to achieve weight transfer with trailed implements known as the Swedish Hitch. This inventive dealership had developed several modifications for MF tractors to make them more

An early MF publicity photograph of the Red Giant line-up with the MF130 at the far end.

The MF Engineering block in Maudslay Road, Coventry, circa 1965.

suitable for forestry work, since for many Swedish farmers forestry contributed a fair proportion of their net income.

So it was at this late 1962 meeting that the Detroit Experimental (DX) Program was ratified, with a caveat that it was to be achieved over a two-year time span. One of the basic requirements of this program was to ensure components manufactured in different countries would be interchangeable, should that become necessary, although there would of course have to be allowances for varying specifications to meet the legal requirements of different countries. Not all this development work would be centred in Detroit; the UK Coventry facility as well as the French Beauvais operation of MF would contribute to the programme as well.

In the UK the MF Engineering Design offices in Maudslay Road, Coventry, were headed by Dr William Willetts (always referred to by the engineering boys as Doc Willetts). The Design offices took up their share of this massive development project when prototypes were ready for field testing. This took place at a number of locations but the Field Test Department was based at a large house at the back of the Banner Lane plant by No.3 gate, in Broad Lane. The manager was Charlie Nicholson and his test team included David Lees and Roger Read. One of the favourite locations for field testing was at Eastnor Castle in Hereford, the home of a great friend of Harry Ferguson, Major Ben Harvey-Bathurst, who also allowed Land Rover Ltd to use his estate for testing their vehicles. Most of the 100 Series tractors on test were painted in Ferguson Light Grey.

Looking briefly at the models originally planned, we find that the French-built MF825 would be upgraded and restyled to become the MF130, the smallest in the range. Next, the well-proven MF35X would be enhanced by a whole raft of new features, and with new sheet-metal styling it would be in line with the other models in the range; it would be designated the MF135 in the UK, in France the MF140, and in America it would be known as the MF150.

Likewise the MF65 MKII would be upgraded and

To give an idea of what the test MF135s would have looked like when painted in Ferguson Grey, here's one standing outside David Mellor's exhibition building at Hathersage near Sheffield.

An aerial view of the Banner Lane site of just over 100 acres. Clearly visible is the tower block built in 1966, which was demolished at 10am on the 8th July 2012.

restyled to become the MF165. A new model would be developed, the MF175, a larger and more powerful version of the MF165. Alongside this, two totally new and larger models, the MF1100 and the MF1130, would be developed, very much with the American market in mind but available to dealers throughout Europe and beyond.

It would be worthwhile quoting directly from EP Neufeld's definitive work *A Global Corporation*, which was commissioned by MF in 1968: "Much of the development work was done by LE Elfes and BF Willetts under JJ Jaeger, who was Vice President of Engineering at MF Detroit. Jaegar, a graduate of the Drexel Institute of Technology and also a graduate of The Massachusetts Institute of Technology, came to MF in 1961. Prior to that he was President and director of The Pratt and Whitney Company Inc. He replaced Herman Klemm, who had plans for an early retirement. [Klemm had been a key player in the development of the MF35 and 65, as described in my book *Massey Ferguson MF35 & 65 In Detail*].

The machine shop at Banner Lane showing transmission casings awaiting their contents.

"An impression of the scale of the work involved in the development of these new tractors is given by a number of new drawings. The MF1100 and MF1130 were completely new so all the drawings were new. The MF135 had 598 changes unique to it, and a further 454 changes in common with some of the bigger tractors. These figures confirm that although the MF135 was indeed based on its immediate predecessor, there was a good deal more to it than simple restyling. The MF175 had 601 new drawings peculiar to it. A further 632 changes were made, common to the smaller models. Over a period of two years the engineering division was allocated one million man-hours to design and build prototypes for testing of the whole range of new tractors, at a cost of $7.5 million. Capital expenditure to manufacture the new range amounted to about $22 million. A further $10 million approximately was required for the development of appropriate diesel engines by the F Perkins arm of MF. Alongside this colossal commitment of resources one must not forget that on 23 January 1959 an enormous amount of money, reckoned to be about US$12.5 million, had been invested in the acquisition of F Perkins of Peterborough as well as the purchase of the Banner Lane facility in Coventry, which was announced on the 23rd July 1959.

"Concurrently with this program of development

The drawing office at Maudslay Road.

MF executives were well aware that Ford was in the throes of developing its own World Wide Range of tractors, known by their management as the 6X Range, to be marketed later as their 1000 Series. These had horsepower outputs from 35 to 65."

The new MF100 Series was launched in the UK in December 1964 at the Smithfield Show and hailed as The Red Giants. At the same event Ford unveiled their 1000 Series, quickly dubbed The Giant Killers! The lasting testament to this human endeavour is the fact that the MF100 Series was in production from late 1964 through to 1979 in various guises, and some of their fundamental features survived in use for much longer.

A Massey Ferguson publicity photograph designed to show the extra length of the Super-Spec model. These featured metal cladding to the cabs, as opposed to the flexible cladding that was generally fitted to the more basic models.

Chapter 1

MF100 Series Tractors Built Worldwide 1964-1988

The main body of this book gives a detailed account of the 100 Series tractors manufactured at Banner Lane, Coventry, with the addition of the small MF130 that was imported from the French MF facility at Beauvais in France. In this first chapter we take a look at most of the models produced by MF and their licensees worldwide over a period of about 23 years. In addition to my own archive material I have relied heavily on John Farnworth's excellent book *A World-Wide Guide to Massey Ferguson 100 and 1000 Tractors 1964 to 1988.* I am most appreciative of John's permission to do so.

The North American tractors were launched in the latter part of 1964, branded "A Rugged New Breed". The range comprised MF135, MF150, MF165, MF175, MF180 and MF1100.

A sales brochure for the French-built MF122 of 1969. The Perkins A4-99 produced 24hp at 2000rpm.

In the UK the 100 Series MF tractors were launched as "The Red Giants" at the Smithfield Show in London in 1964, but were not available to customers until January the following year. The range initially marketed in the UK comprised the following models: MF130, MF135, MF165 and MF175.

In France, like the UK, the 100 Series tractors did not come on to the market until early January 1965, and the range offered to French farmers consisted of MF122, MF130, MF135, MF140, MF145 and MF165.

The Italian-built tractors in the 100 Series were first marketed in 1969, with three crawler tractors introduced around the same time; MF100 Series tractors continued to be produced in Italy until 1988!

It was not until 1974 that German-built MF100 Series tractors became available.

As John Farnworth points out, the MF associated companies in Mexico and Morocco were making some 100 Series tractors by the late 1960s.

In this summary we will work through the range in numerical order, setting out the basic features and salient dimensions model by model. Please note that the tyre sizes quoted are those given in MF publications of the time; as a result there are differences in the way they are expressed.

MF122

Smallest in the range was the French-built MF122 Standard, powered by a Perkins A4-99 indirect-injection engine (Perkins-patented "Howard Chamber" pre-combustion type) producing a quoted 24bhp at 2000rpm and driving through a two-stage clutch to an eight-speed forward and two-speed reverse gearbox. Differential lock was standard and the brakes were of the dry disc type. Tyre sizes were front 4.50 x 16, rear 9.00 x 28, weight 1170kg, and overall width 1600mm at minimum track setting. The MF122 Vineyard was basically the same machine but the tyre sizes were reduced to 4.00 x 5 front, 9.0 x 24 rear, and overall width was 890mm at minimum track setting.

MF125

Next in the numerical line-up is an oddball, the Japanese MF125. This model was produced in 1973 by Mitsubishi Heavy Industries at the Sagamihara works especially for the Japanese market under licence from MF. It was powered by Mitsubishi's own KE135 diesel engine, a two-cylinder four-stroke watercooled unit with an output of 25bhp.

MF130

Then we come to the MF130 produced in France from 1965 at the Beauvais factory. These tractors were sold in several variants throughout Europe. A detailed outline of the MF130 is the subject of the next chapter.

MF132

A latecomer to the 100 Series was the MF132 produced in Germany in 1975, not only in standard width but also as a narrow version and a 4WD model. The tractors were powered by an Eicher EDK2 aircooled two-cylinder diesel engine developing 34bhp at 2150rpm, driving through a single-plate clutch to a six-speed forward and two-speed reverse speed gearbox. The brakes were of the drum type. The hydraulic system offered Draft and Position Control as well as external services. Tyre sizes were 4.00 x 12 or 4.50 x 16 front, 7.50 x 18 or 8.00 x 24 rear. Tyre size for the front wheels of the 4WD model was 7.00 x 12 or 6.00 x 16, while for the rear wheels there were two options, 7.50 x 18 or 8.00 x 24. The overall width of the 4WD version was 920mm, and 880mm for the narrow model.

MF133

The French-produced MF133 was made in Standard and Vineyard widths. These tractors were powered by the direct-injection Perkins AD3-144 engine, similar to but of smaller capacity (2.36 litres) than the more widely known AD3-152 engine, which had a capacity of 2.5 litres. The clutch was of the dual type, feeding power into an eight-speed forward and two-speed reverse gearbox. Drum brakes were fitted at the wheel hubs. Front tyre size was either 6.00 x 16 or 4.00 x 19, while the rears were 12.4/11 x 28 or 11.2/10 x 28. The overall width was 1620mm and weight

MF130 brochure.

MF133 photographed in Germany, but built at Beauvais in France, fitted with a Perkins A3-144 engine.

The driver's view of the dash panel of the MF133. Note the raised air intake stack and, to the left of the steering wheel, direction indicator switch and warning lights.

1440kg. The hydraulic system offered Draft, Position and Response Control and external services. A later French incarnation from the 1970s was the MF133 Super, whose specification was similar to the model just outlined but with the option of being fitted with Category 2 lower links.

MF134V

A much later model in the range, in 1986/87, was the MF134V produced in Italy in 2WD and 4WD forms. Both were powered by the 2.5-litre Perkins A3-152 diesel engine. A single-plate clutch connected to a twelve-speed forward and four-speed reverse gearbox. The brakes were triple disc type. Tyre sizes for the 2WD model were front 7.00 x 12, and rear 12.4 x 24. The 4WD models were equipped with 7.50 x 15 front and 12.4 x 24 rear tyres. Bare weight was quoted as 2100kg for the 2WD and 2175kg for the 4WD, both models having the same width of 1130mm. The MF hydraulics had Category 1 attachment pins.

MF135

The MF135 was launched late in 1964. Variations of this model were built in North America, France and the UK. The UK-built tractors are the subject of Chapter 3. As for the American versions, two power units were available, either the Perkins A3-152 direct-injection diesel engine or the Perkins petrol engine designated the AG3-152. The "152" in both designations signified that both engines had the same capacity of 152.7cu. in (2.5 litres). The diesel produced a quoted 45.5bhp at 2250rpm while the petrol version was quoted at 35bhp. Some early North American MF135s were built with the Continental Z142 petrol engine, and a few economy models (economy in terms of price) were powered by the Continental Z-134 petrol engine, the same unit as was fitted to some TO35 tractors! The North American-made tractors were fitted with lighting equipment and had the headlamps mounted on the sides of the bonnet panels, unlike the British and French 135s which had the headlights mounted within the radiator grille.

Generally speaking the weight of the North American 135s was of the order of 1420kg (3130lb) depending on the model, and the overall width 64in (1625mm). Like the UK, the Americans produced a Vineyard model available with either diesel or petrol engine. They also offered users the MF135 Orchard Special, again with a choice of fuel options. The main difference in its specification was in the tyre sizes, which were 7.50 x 10 front and 18.4/15 x 16 rear.

French-made MF135s were produced in Standard, Narrow and Vineyard formats. All were fitted with the Perkins A3-144 engine, which had a bore and stroke of 88.9 x 127mm, giving a capacity of 2360cc. The compression ratio was 16.5:1. Generally speaking, the Beauvais-built MF135s were very similar to those produced at Banner Lane, albeit with this slightly smaller version of the Perkins engine. At a later stage,

A US-built MF135 powered by the Continental Z142 gasoline engine.

An MF135, with outboard rectangular headlamps and a front bumper, harvesting apples in Italy.

in 1972, the MF135 Super was introduced. It had flat top rear fenders on the Standard and Narrow models but the Vineyard version was fitted with the shell-type fenders. These later models were fitted with an eight-speed forward and two-speed reverse gearbox. The French-built Vineyards were only available with tyre sizes of 5.00 x 15 or 5.50 x 15 front and 9.00 x 28 or 10.00 x 28 rear. By 1980 the MF135 Mk III was being produced at Beauvais, now powered by the Perkins A3-152 diesel engine. These had a dual clutch, and the eight-speed forward and two-speed reverse gearbox had synchromesh between the four ratios of the main box. These later models had oil-immersed five-plate disc brakes.

In 1980 Ebro, the Spanish licensee (Ebro is the name of a Spanish river), was producing Narrow and Vineyard versions of the MF135, again powered by the Perkins A3-152 diesel engine. Generally these tractors were fitted with a single-plate clutch, but a dual type was available as an optional extra. The gearbox had the usual six forward speeds and two reverse; available as an accessory was a side-mounted PTO. Brakes were drum-type, while the hydraulics gave the operator a choice of Draft, Position and Response Control. If pressure control was needed it could be specified as an extra-cost factory fitment. Front tyre sizes available on both models were 5.50 x 16 or 6.00 x 16. Rear tyre sizes offered on the Narrow model were either 10 x 28 or 11 x 28. For the Vineyard tractor the choice was either 11 x 28 or 12 x 28. Both models had the same overall length of 3000mm, the widths being 1370mm and 1200mm.

A French-built MF135, selling garlic at the roadside.

Between 1972 and 1976 Ebro were producing a standard-width MF135 with a similar specification to those just outlined but with the option of six- or eight-speed gearboxes. The Ebro built MF135 had a "through the bonnet" air intake stack terminating with a centrifugal bowl type air pre-cleaner.

John Farnworth's book goes on to give details of the French produced MF135 Frutteto of 1981. This model was fitted with the Perkins A3-152 diesel engine coupled to an eight-speed forward and two-speed reverse gearbox. The brakes were of the oil-immersed disc type and the hydraulics offered the usual functions. Front tyre size was 5.00 x 15, and rear either 10 x 24 or 10 x 28.

Another Continental-powered, US-built MF135; this 1964 example is set up for use on the golf course.

MF135S

It may be appropriate here to mention that in 1977 MF135s produced at Coventry for export to the Swedish market were badged MF135S. This model was built to the usual UK specifications but had fitted as standard power-assisted steering as well as a cab, with the upper part painted silver, incorporating a heater and windscreen defroster.

MF139 &139A

By 1980 the MF German facility were manufacturing their MF139 and 139A tractors, with two- and four-wheel drive respectively. These were powered by an Eicher EDK3-4 three-cylinder air-cooled diesel engine with a quoted power output of 40bhp DIN at 2150rpm, coupled via a dual clutch to an eight-speed forward and two-speed reverse gearbox. The brakes were of the drum type. The hydraulic system featured the usual Draft, Position and Response Control plus a floating facility. Tyre sizes for the 2WD tractors were front 8.00 x 16, rear 9.50 x 24. The 4WD machines had tyre sizes of 6.00 x 16 front and 9.50 x 24 rear. These tractors weighed in at 1555kg for the 2WD and 1715kg for the 4WD.

MF140

The French-built MF140 came on to the market in 1965. It was powered by the Perkins A3-152 diesel engine and had a dual clutch feeding power into a six-speed forward and two-speed reverse gearbox, compounded by a High/Low unit, with the MF eight-speed box as an alternative. Multi-Power could be supplied as an optional extra but only with the six-speed gearbox. Brakes were of the drum type and a centre PTO could be specified. The hydraulics were the usual: Draft, Position and Response Control, with provision for external services. Tyre sizes were front 6.00 x 16, rear 11.00 x 28. Track setting was front 1220-2030mm, rear 1220-1930mm. Overall width was generally 1620mm, and weight 1450kg. Interestingly, the Standard models could be ordered with a passenger seat at extra cost. Concurrently with the Standard model a Narrow MF140 was available with an overall width of 1370mm, weighing 1400kg. Tyre sizes were 4.00 x 19 or 5.00 x 16 front and 10.00 x 28 or 11.00 x 28 rear. MF in France naturally produced a Vineyard version in the 140 range; all were fitted with 28in rear tyres. By about 1970 this whole range had been upgraded to become the MF140 Super, the principal changes being the introduction of a straight front axle and flat-top fenders. As an optional extra a 4:1 reduction sandwich gearbox was made available. The passenger seat remained an accessory but was only suitable for the Standard tractors.

MF142 & 142A

By 1975 the German MF factory were producing their MF142 in both 2WD and 4WD. They were powered by an Eicher EDK3-3 three-cylinder direct-injection air-cooled engine with a quoted output of 52bhp DIN at 2150 rpm. A dual clutch was fitted, mated to an eight-speed forward and two-speed reverse gearbox. The brakes were of the drum type, whilst the hydraulic system provided Draft and Position Control, external services and a floating facility. The 2WD models were equipped with 5.00 x 16 front and 10.00 x 24 rear tyres. On the 4WD version the tyre sizes available were front 6.00 x 16 or 7.00 x 18, and rears 12.4/10 x 24 or 12.4/11 x 24. The overall width of both variants was the same at 1010mm, but the length of the 2WD was 2870mm and the 4WD 2950mm. Both models had differential lock as standard, the 4WD tractors having differential lock on both axles. The German factory also produced a Narrow model in this range, built more or less to the same specification. By 1978 the 142 was upgraded to become the MF142/142A MK II, again in two- and four-wheel drive format. Steering was still unassisted on the 2WD but hydrostatic on the 4WD. The hydraulic pump now operated at a higher pressure and mention is made that the lower links could be specified as Category 1, 2 or 3! It seems a bit strange that Category 3 was offered.

A 1969 French brochure of the MF140 which was powered by a Perkins A3-152 engine.

MF145

The French-built MF145 was introduced in 1965 and was very similar to the UK-built MF135, which is covered fully in Chapter 3. It was fitted with a 'through the bonnet' raised air intake stack with pre-cleaner. Also as an extra a conversion kit was made available to give either Category 1 or 2 linkage attachment. A narrow model was also built within this range, as was the Vineyard, the MF145V, which had the same basic specification but with an eight-speed gearbox as well as oil-cooled three-plate disc brakes; overall width was down to 1050mm. Tyre sizes offered were front 6.00 x 16 or 5.50 x 16, rears 11.2/10 x 28.

MF 147 Compact

Offered by the Spanish arm of MF in 1980, this model was again powered by the ubiquitous Perkins AD3-152 engine, with drive taken through a dual clutch to the eight-speed forward and two reverse gearbox; the brakes were of the drum type. The hydraulics had the usual Draft, Position and Response Control with external services. A side PTO was an optional accessory.

MF148

Introduced by MF in 1972, this was manufactured both at Coventry and Beauvais to very similar specifications. The UK-built MF148 is dealt with in Chapter 4. It should be indicated here that the Beauvais factory did produce a MF148 MK III Super Narrow starting in the late 1970s. Again this was to the same basic specification but only the eight-speed transmission was available, with dual clutch. Track settings were 1040-1450mm front and rear 970-1520mm, weight 1040kg.

MF150

The American range of MF 150s was manufactured in three different configurations: Standard, Wide Row Crop and Tricycle, with the choice of diesel or petrol engines, these being the Perkins diesel AD3-152 or the Continental Motors petrol Z145 which, with the MF takeover of Perkins in Peterborough, was later replaced by the Perkins AG3-152 petrol engine. The standard gearbox was the MF six-speed forward and two reverse, with Multi-Power offered as an optional extra. The Standard and petrol Row Crop had hub-mounted drum brakes, while the diesel Row Crop had double dry disc brakes mounted within the trumpet housings. There was a dual clutch on the earlier models, but on later from 1972 models a split-torque clutch was available as an alternative. The hydraulic system offered the usual Draft, Position, Response and Pressure

A US-built MF150 imported to the UK by Julie Browning and Peter Smith. Note the spring suspension seat and Power Adjustable Variable Track (PAVT) rear wheels.

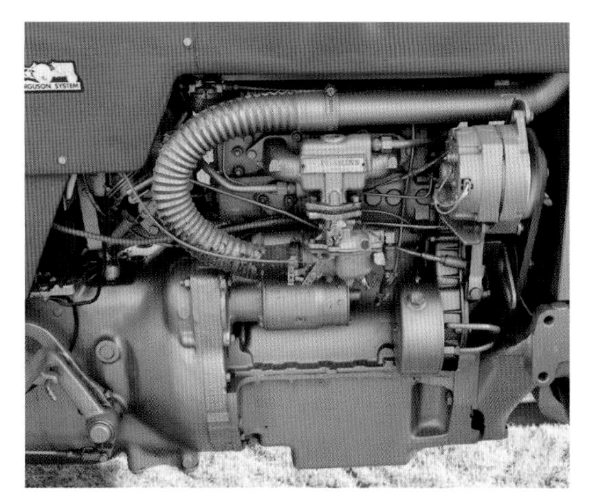

It is powered by a Perkins AG3-152 petrol engine.

Control together with external service tappings. Differential lock was an optional fitment as was power steering. Tyre sizes were: Standard, front 6.00 x16, rear 13.6/12 x 28; Wide Row Crop, front 6.00 x 16, rear 12.4/11 x 38; Tricycle, front 5.50 x 16, rear 12.4/11 x 38. Normal track setting for the Standard model was 1220mm, and 1270mm for the Wide Row Crop model. Weights (with diesel engines) were Standard 1735kg, Wide Row Crop 1826kg, Tricycle 1780kg.

By 1972 MF in Mexico were building a MF150 in both Standard and High Clearance forms. In general terms these tractors were similar to those built in North America but the headlamps were built into the radiator grille like their counterparts the UK MF135 or the French-built MF145. Interestingly, the rear fenders followed a similar design to those of the North American MF150. The specification comprised the Perkins AD3-152 engine, a dual clutch, a six-speed forward and two reverse gearbox, and double dry disc brakes. The hydraulic system offered Draft, Position and Pressure Control as well as external services. Tyre sizes were 6.50 x 16 front and for the rear a choice was available, either 13.6/12 x 38 or 14.9/13 x 28. Standard equipment included manual steering, differential lock, lights, front weights and weight frame – also a cigarette lighter, no doubt a cultural necessity! A High Clearance version was also available. Weights were Standard 2180kg, High Clearance 2320kg.

From 1972 through to 1976 the MF Argentina factory offered an MF150 Vineyard. This model was again powered by the Perkins AD3-152 engine, driving through a dual clutch to an eight-speed forward and two reverse gearbox. The brakes were double dry disc of 222mm diameter. The earlier models in this range had a hydraulic system that featured Draft and Position Control plus external services, while on later tractors Response Control was added. The air cleaner was the oil bath type with a stack 'through the bonnet' towards the front, equipped with a pre-cleaner. Brakes were double dry disc of 222mm diameter. Only manual steering was available. Overall width was 1250-1450mm. Tyres were front 5.00 x 15, rear 12.4/11 x 28.

The French-built MF152 was powered by the Perkins AD3-152 engine. Note the sandwich reduction gearbox and the three-piece rear wheels on this example.

MF152

By 1972 Beauvais were building their MF152 in Normal and Narrow forms. Again these tractors were powered by the Perkins AD3-152 engine, with two transmission options: dual dry clutch or single dry plate split-torque type. The choice of gearbox was either eight-speed forward and two reverse, or the twelve-speed forward and four reverse with Multi-Power. The brakes were five-plate oil-cooled discs. Earlier models in this range had hydraulic systems with just Draft, Position and Pressure Control plus external services, but later models incorporated Response Control as well. Both Category 1 and 2 linkages were available. This model was listed in the July 1979 dealers' Pocket Catalogue as being available on the UK dealer network, and was still listed in a December 1985 issue of the Pocket Catalogue (the oldest I have in my archive). By 1980 MF in France were producing an MF152 Mk III Orchard model, yet again powered by the Perkins AD3-152 engine fitted with a dry dual-element air filter. The engine output was fed through a dual clutch to an eight-speed forward and two reverse gearbox with synchromesh on the four forward ratios in the main box. The hydraulic system had Draft, Position and Response Control as well as external services. The brakes were of the three-plate oil-immersed type. Tyre sizes were front 6.00 x 16 and rear 12.4/11 x 28 or 13.6/12 x 28, with cast centre PAVT wheels. As optional equipment a 4:1 reduction sandwich-type gearbox was made available as was a manual reverse shuttle. Width overall was 1350mm and weight 1800kg. A 1983 salesman's Pocket Catalogue lists the French-made MF152 Mk III Narrow as being sold in the UK as well as in France. The specification generally follows that of the previously outlined model with the exception of a narrow overall width, pressed steel wheels all round, and tyre sizes of 6.00 x 16 front and 11.2/10 x 28 rear, with weight at 1660kg. Yet another variation was the French MF152 Mk III Super Narrow, again similar in specification to the one just mentioned but with its overall width further reduced to 1250mm. Also listed in a 1983 Pocket Catalogue was the MF152S Special Tractor, again a product of Beauvais but marketed in the UK. The caption reads "A tractor that is invaluable to farmers who need versatility combined with minimum width. Ideal for a variety of special applications. Fitted with an easily removable safety frame". Overall width is quoted as 1245mm on a 1016mm track setting.

MF154

The 1975 Italian-built MF154's specification included the Perkins AD3-152 engine and a dual clutch driving a twelve-speed forward and four reverse gearbox with constant mesh and some synchromesh. The hydraulics gave Draft and Position Control plus external services. With Category 2 linkage, these tractors had their hydraulic pump mounted directly on the engine. The sheet metalwork design was unique to the Italian tractors. Tyre sizes were front 6.00 x 16 and rear 13.6/12 x 28. Weight was 2080kg. Track was adjustable at the front from1300-1800mm, and at the rear from 1200-1900mm.

Manufactured alongside this model was a 4WD version, the MF154-4, built to the same basic specification with the exception of the front tyres being 9.50 x 20. The front axle hubs incorporated planetary reduction gears. The manually-engaged central drive shaft to the front axle followed a design patented by Same in 1966. A front differential lock was provided as standard, mechanically engaged, as was the rear one, and as an extra-cost option the front lock could work automatically. Tyre sizes were 2WD front 5.00 x 16 or 5.50 x 16, 4WD front 6.50 x 16 or 7.50 x 18. Rear tyres were 2WD 11.2/10 x 24 or 12.4/11 x 28, 4WD 11.2/10 x 24 or12.4/11 x 28. Width of both two- and four-wheel drive models was the same at 1264mm, weights being 2050kg and 2190kg respectively.

Two other variants of this range were produced in Italy, again both in two- and four-wheel drive versions: the MF154V, a Vineyard tractor with an overall width of only 980mm, and the MF154F, again available in two- or four-wheel drive, this fruit tractor having a width of 1490mm.

MF155

In the early 1970s the Beauvais factory was manufacturing the MF155 Standard and Narrow. These tractors were powered by a Perkins AD4-203 direct-injection diesel engine with a quoted power output of 54bhp DIN at 2000rpm (the same engine as fitted to the Coventry-built MF65 Mk II). The standard transmission was an eight-speed forward and two reverse gearbox with a two-stage dry clutch. An alternative build was a single dry plate split-torque clutch coupled to a six-speed forward and two reverse gearbox with Multi-Power to double up the ratios available – with this arrangement the PTO was truly independent, with its own hydraulic clutch with built-in brake. A side PTO was listed as an optional fitment. The braking was provided by five-plate oil-immersed 222x188mm discs. The hydraulic system offered Draft, Position and Pressure Control. Two hydraulic pumps were standard. The linkage pump gave an output of 14.1 litres/min whilst the auxiliary pump gave 28.6 litres/min. Tyre sizes offered were front 6.00 x 19 and rear 12.4/11 x 36, 13.00 x 28 or 16.9/14 x 28. Overall width on 6.00 x 19 front tyres was 1830mm with a track setting of 1420mm. The Narrow model had tyre sizes of 6.00 x 16 front and 12.00 x 28 rear, with overall width reduced to 1440mm.

In 1972 the Argentinean MF factory was producing MF155s in Standard and High Clearance form. These models were powered by the faithful Perkins AD3-152 producing 45hp DIN at 2000rpm. A dual clutch was fitted, driving an eight-speed forward and two reverse gearbox. The brakes were the double dry disc type of 222mm diameter. The hydraulic system offered Draft, Position and external services control; the later tractors had Response Control added. Tyre sizes were front 6.00 x 16 or 6.50 x 16, and rear 12.4/11 x 28, 13.6/12 x 38 or 14.9/13 x 26. Weight quoted for the Standard model was 2250kg.

A Spanish specification sheet from Motor Iberica SA dated 1970 shows the company's EBRO MF155D, available in Standard and Narrow widths. These tractors were in reality Fords with MF styling. The engine is described as an EBRO unit but although made by EBRO these tractors were based on designs and tooling purchased from the Ford Motor Company, so in fact the whole skid unit was a Fordson Super Major dressed up with MF-style sheet metalwork. The engine was a four-cylinder direct-injection unit

The MF155 was powered by the Perkins AD4-203 – the same engine as fitted to the MF65 MkII.

The Ebro MF155D was available in standard and narrow widths. It was in fact a Ford with Massey Ferguson styling.

of 3.6 litres producing a quoted 55bhp at 1800rpm. The clutch was single-plate and PTO was optional. The gearbox had six forward and two reverse speeds. The hydraulics, obviously by Ford, gave Qualitrol (Ford's name for MF's Draft Control) and Position Control, with Category 2 linkage. Tyre sizes for the standard model were front 7.00 x 16, and rear 11.00 x 36, 12.00 x 36, 12.00 x 38 or 14.00 x 30. For the Narrow tractor the sizes were front 6.00 x 16 and rear 11.00 x 28 or 13.00 x 24. Widths were maximum 2160mm, minimum 1650mm for the Standard, and for the Narrow 1430mm. The respective weights were 2300kg and 2170kg.

A French-made MF158 MkIII collecting fallen apples.

MF157

This was produced by EBRO in 1972. Its brief specification shows it to be a true MF design, being powered by the Perkins AD4-203 direct-injection diesel engine with a quoted 62bhp at 2250rpm. (The same Perkins engine in the Mk II MF65 built in Coventry produced a quoted 58bhp at 2000rpm). No details of the gearbox or brakes are to hand. A dual clutch was fitted. The hydraulic linkage had Category 2 attachment points and offered Draft, Position and Pressure Control. Normal width was 1830mm. Tyre sizes were front 6.50 x 16, rear 12.00 x 36. A sprung seat was standard, the air intake protruded through the rear of the bonnet – and there was a cigarette lighter for driver convenience!

MF158

This model was produced at Beauvais in both Standard and Narrow versions around 1972-1976. The tractors had the Perkins AD4-203 engine with a quoted power output of 56bhp at 2000rpm. The air cleaner was a dry single-element type. For the transmission two systems were available, the standard being a single-plate split-torque clutch mated to an eight-speed forward and two reverse gearbox, thus giving a truly independent PTO. Alternatively and at extra cost tractors could be supplied to customers with a dual clutch mated to MF's 12-speed gearbox incorporating Multi-Power. The brakes were of the oil-immersed five-plate type. The hydraulic system offered Draft, Position and Pressure Control with external services; later tractors in this group were equipped with Response Control as well. Tyre sizes on the Standard-width model were front 6.00 x 19, rear 12.4/11 x 36, 14.9/13 x 28 or 16.9/14 x 26, overall width was 1830mm, and weight 2000kg approximately. The French factory continued to produce a variety of tractors in the MF158 Mk III group through to 1986. To simplify the text in this area I will list them model by model, pointing out salient differences rather than repeating the whole build specification. They were all powered by the Perkins AD4-203 diesel engine, all had eight-speed forward and two reverse gearboxes with synchromesh, and the clutch was of the split-torque type. The hydraulic system had the usual MF features, the fuel tanks held 68 litres, and the brakes were three-plate oil-immersed.

- MF158 Mk III Wide Vineyard. Tyres were front 6.00 x 16, rear with cast centre PAVT wheels 13.6/12 x 28 or 14.9/13 x 26. Weight was 1865kg and overall width 1260mm.
- MF158 Mk III Orchard. Tyre sizes were front 6.00 x 16, rear with cast centre PAVT wheels 13.6/12 x 28

or 14.9/13 x 28. Weight 1900kg and overall width 1396mm.

- MF158 Mk III Narrow. Tyre sizes were front 6.00 x 16, rear 13.6/12 x 28, Category 1 and 2. Width overall 1340mm, weight 1865kg.
- MF158 Fruit Tractor. Power steering as standard, front tyres 6.00 x 16, rear on cast centre PAVT wheels 13.6/12 x 28, weight 2045kg, width 1390mm.
- MF158V 1982. Category 1 hydraulics. Tyre sizes front 4.50 x 16, rear 12.4/11 x 28

MF160

The EBRO MF160 was manufactured by the Spanish firm of Motor Iberica. As mentioned when outlining the EBRO MF155, those tractors as well as the ones we are looking at now were in reality a Ford tractor design finished off with MF-style sheet metalwork. In fact the EBRO MF160 was in many ways very similar to the EBRO MF155, but with the same basic engine being allowed to run a bit faster and thus giving a slightly increased power output of 61bhp at 2100rpm. The hydraulic systems were all of Ford design. Tyre sizes were 6.00 x 19 front, and 12 x 36 or 14 x 30 rear. Weight was 2400kg, and width overall between 1650mm and 2160mm. This information is provided by John Farnworth's specification sheet dated 1966. A later 1970 sheet of his lists an EBRO MF160D in Standard and Narrow widths. The same skid unit of Ford design was used as in the EBRO MF155. Tyre sizes for the Standard model were 7.00 x 16 front, 12 x 36 or 14 x 30 rear, and for the Narrow model the sizes were 6.00 x 16 front, 11 x 28 or 13 x 24 rear .Weights were 2400kg and 2350kg respectively.

MF164 & MF165 (Italy)

The Italian factory was producing several models in its MF164 range in the latter part of the 1980s. The MF165S was produced in both two- and four-wheel drive forms. The engine chosen was the Perkins AT3-1524 (T indicating that it was turbocharged and the suffix 4 denoting a stage of that engine's development). This engine had a quoted output of 53bhp DIN at 2250rpm. The clutch was a dual type, the gearbox gave twelve forward and four reverse speeds, and a creeper box could be specified to give twenty forward ratios. Hydraulics were powered by an engine-mounted gear pump to provide Draft, Position, Response and Intermix Control. The linkage attachment points could be either Category 1 or 2. Rear differential lock was standard on both 2WD and 4WD models but an optional automatic differential lock could be specified for the front axle of the 4WD models. The PTO offered 540rpm and 1000rpm at engine speeds of 1944rpm and 1916rpm respectively. The brakes were oil-immersed discs, hydraulically activated. Tyre sizes for the 2WD were front 5.00 x 15, rear 11.2/10 x 24, and for the 4WD models, front 6.50 x 16 or 7.50 x 18, rear 11.2/10 x 24 or 12.4/11 x 28. The minimum width of both models was 1485mm, weights being 2100kg 2WD and 2200kg 4WD.

Also produced by the Italian factory, again in the late 1980s, was the MF164V, produced in both two- and four-wheel drive versions. These tractors had basically the same specification as the model just mentioned. The hydraulic linkage differed in that it was only available as Category 1. Tyre sizes were 2WD front 7.00 x 12, rear 12.4/11 x 24. The 4WD tractor was equipped with 7.50 x 16 front and 12.4/11 x 24 rear tyres. The weights were 2100kg and 2175kg respectively and both were 1130mm wide.

The final model in this Italian range at this time was designated MF164F, basically the same build as the others but 1330mm wide – more of a fruit tractor!

MF165 (Britain)

This model, launched in 1964, is the subject of Chapter 5.

MF165 (France)

The MF factory at Beauvais manufactured MF165s that more or less followed the specification of those built at Coventry, but by 1970 they were offering MF165s as a Mk III version fitted with the more powerful Perkins AD4-212 engine. These Mk IIIs were available in three different heights: Standard with a ground clearance of 350mm, Medium with a ground clearance of 400mm, and High Clearance with a clearance of 480mm. Tyre sizes were: Standard tractor, front 6.00 x 16, rear 13 x 28; Medium, front 6.00 x 19, rear 11 x 36, 12 x 36 or 14 x 30; High Clearance, front 6.00 x 16, rear 12 x 38.

MF165 (USA)

Turning now to the MF165s built at Detroit and introduced to the market late in 1964, we find that there were four types produced, the Standard model, a High Clearance, a Row Crop and a Dual Wheel Tricycle. Being North American, both diesel and petrol engine options were made available to customers. The engine options offered were either the Perkins diesel AD4-203 as used in the early UK-built 165s (which are described in Chapter 5) or the Perkins petrol AG4-176 (176cu.in.) on earlier models. By 1970 the petrol engine installed was the Perkins AG4-212, which delivered 51.9bhp at the PTO – by comparison the diesel AD4-203 delivered 52.4bhp at the PTO. The clutch was either a dual type or a single split-torque unit for a truly independent PTO. The gearbox was the usual MF six-speed forward and two reverse, but Multi-Power could be specified. Later the option of MF's eight-speed gearbox was offered, according to an American 1973 MF Machinery Digest. Brakes were double dry disc of 7in diameter. The hydraulic system offered Category 1 and 2 linkage, and the lift pump operated to 3000psi with a delivery rate of 4.8gal/min, with the usual MF functions of Draft, Position and Pressure Control with external services. To increase hydraulic output an auxiliary pump could be installed which boosted the flow rate to 12.8gal/min. The Standard model had ground clearance of 14.75in, the Row Crop model 19.3in, the Low Profile 12.5in. The Dual Wheel Tricycle was only available with a diesel engine and had a ground clearance of 12.5in. On all models the rear wheels were fitted with cast centres of the PAVT type.

MF165 (Argentina)

In Argentina in 1972, the MF165s were being manufactured in Standard and High Clearance. The engine again was the faithful Perkins AD4-203 diesel, a dual clutch delivered power to the MF eight-speed forward and two reverse gearbox, and the brakes again were double dry disc. Hydraulics were normal MF type with Category 2 linkage. Power-assisted steering was optional and the usual range of tyre sizes was offered. By 1975 the Argentinean facility was producing MF165s fitted with the Perkins AD4-236 engine (236cu.in/3.87 litres). These engines produced 69bhp at 2000rpm, or at the PTO 66bhp at 540rpm. A dual clutch was fitted, mated to an eight-speed forward and two reverse gearbox. Tyre sizes were front 7.00 x 16, and rear 16.9/14 x 30, 13.6/12 x 36, 13.6/12 x 38 or 18.4/15 x 30. Overall width was1870mm and weight with 16.9/14 rear tyres was 2200kg.

MF165 (Spain)

EBRO Motor Iberica of Spain produced MF165s powered by a Perkins A4-236, this time producing 70bhp at 2000rpm (working bhp 66). A dual clutch was fitted, there were double dry disc brakes, and a front-mounted 'through the bonnet' air intake stack with a plain gauze filter was standard.

MF165 (Sweden & Denmark)

In Sweden MF marketed models of the MF165 built to certain different legal and climatic requirements. These were built at Coventry and then exported. The earlier tractors were powered by the Perkins AD4-203 while later models were equipped with the Perkins AD4-212. Some of the tractors exported to Sweden were badged Transport, perhaps as a semi-industrial model, and photographs show them fitted with front mudguards. All were fitted with well heated safety cabs. John Farnworth notes that later (1975) MF165s exported to Sweden had the letter 'S' incorporated into the large circular badge fitted either side of the bonnet. UK-built MF165s exported to Denmark had an 'X' medallion mounted on each side of the bonnet just to the rear of the normal badge and just above the silver bonnet band. On later models a creeper sandwich gearbox was available as an extra; it would have lengthened the wheelbase by 6in (150mm).

MF168 (France)

The MF168 is the subject of Chapter 6, but here we will look briefly at the many variants produced at Beauvais. In general terms their basic build specification and options available followed closely those of 168s produced at Banner Lane, so they were powered by the Perkins AD4-236 engine rated at 66bhp DIN at 2000rpm. The model range included a High Clearance model, as well as a four-wheel-drive model, possibly a first for MF. It had power steering as standard. The front axle had epicyclic reduction gears in the hubs, the drive to which could be engaged or disengaged on the move via a hydraulic clutch pack within the drive train.

The front tyre size was 9.50 x 24, rear 16.9/14 x 30, and the 4WD model weighed in at 2800kg, about 600kg heavier than the standard tractors.

By 1980 the French factory was producing Orchard and Wide Vineyard versions of the 165. They had the same Perkins AD4-236 engines and a single-plate 12in (305mm) clutch coupled to an eight-speed forward and two reverse gearbox, with synchromesh on the forward gears. The PTO was truly independent and controlled by its own hydraulic clutch, and the brakes were oil-immersed five-plate disc type. The Orchard model was shod with 7.50 x 16 front tyres while the rears were either 13.6 x 28 or 14.9 x 28. For the Wide Vineyard version tyre sizes were front 6.00 x 16, rears 13.6 x 28 or 14.9 x 28. The Wide Vineyard had an overall width of 1260mm while the Orchard tractor was 1396mm wide. As optional equipment either a creeper gearbox with a ratio of 4:1 or a manual reverse shuttle could be fitted. Both were of the sandwich type, installed behind the main gearbox. These models had similar sheet metalwork to the earlier models but the radiator grilles were finished in black.

Around 1982 the Beauvais plant was producing a tractor designated MF168S Narrow, with a minimum width of 1290mm and tyre sizes of 6.00 x 16 front,

14.9/13 x 28 rear. This model along with a few others was equipped with power steering as a standard fitment, along with epicyclic reduction hubs on the rear axle. Another model of this era was the MF168S Orchard, 1380mm wide on 7.50 x 16 front and 14.9/13 x 28 rear tyres, again with power steering and epicyclic reduction rear hubs. Yet another very similar model was the MF168 Mk III Frutteto. These tractors had optional tyre sizes, front 6.00 x 16 or 7.50 x 16, rear 13 x 28, 13.5 x 24 or 14.9 x 28. Minimum width was 1400mm, and weight 2260kg. Beauvais also manufactured a model designated MF168P, again very similar, with weight 2290kg, width 1440mm, and tyre sizes 6.00 x 16 front and 12 x 28 rear. John Farnworth also lists an MF168F/4, a four-whee-drive tractor with 9.50 x 20 front and 14.9 x 28 rear tyres. Minimum width is quoted as 1400mm, and weight 2600kg.

MF174

The MF174 was produced in Italy from around 1975. It had different sheet metal styling from most of the others in the 100 Series and was powered by the Perkins AD4-212 engine rated at 61bhp DIN at 2200rpm. The power was fed through a dual clutch giving a live PTO which could be selected to drive relative to engine speed or ground speed. An unusual feature was the provision of a second PTO shaft, terminating above the normal one, which gave direct drive from the engine, presumably with some sort of clutch in the driveline, for belt pulley work. The gearbox was of the constant-mesh type with some synchromesh facility and gave twelve forward and four reverse gears. The hydraulic pump was of the gear type, driven directly off the engine. Brakes were double dry disc. Tyre sizes were 7.50 x 16 front, 16.9/14 x 30 rear. The 4WD model was badged MF174/4 and its basic specification was the same but the front tyres were 9.5/9 x 24 and rear 14.9/13 x 30. The front differential was centrally mounted and the wheel hubs housed epicyclic reduction gears.

By 1988 the Italian factory had upgraded the two models just described to the MF174S, again produced in two- and four-wheel drive. The engine chosen was the Perkins AD-4.23, producing 67bhp DIN at 2200rpm. A dual clutch was fitted giving live PTO. The standard gearbox provided twelve forward and four reverse ratios but a creeper gearbox could be installed giving twenty forward and eight reverse speeds. On the 4WD model power to the front axle was engaged mechanically, but at extra cost a hydraulically operated control could be installed. With the introduction of these upgrades the opportunity was taken to drastically restyle the sheet metalwork, in a bold and individual design statement carried out with typical Italian flair. A variant of this model was the MF174V, the Vineyard, which was 1185mm wide whereas the MF174S was 1358mm wide in both 2WD and 4WD forms. The Vineyard model was also produced in two- or four-wheel drive, and oil-immersed brakes were fitted.

According to John Farnworth an MF174F model was produced with the same basic specification but 1490mm wide for both two- and four-wheel drive versions. Tyre sizes were front 7.50 x 16, rear 13.6/12 x 28 or 14.9/13 x 24 on 2WD tractors; for the 4WD models the sizes were 7.50 x 20 or 7.50 x 18 front, 13.6/12 x 28 or 14.9/13 x 24 rears.

MF175 (USA)

The Coventry-built 175 and 178 are dealt with in detail later in this book, so here we will consider briefly the range of models built under this heading in North America. Their MF175 models were available as Standard, Row Crop and Low Profile. They were marketed with two engine options, both by Perkins: the AD4-236 diesel or the AG4-236 petrol. Both power outputs were the same at 63bhp at the PTO. For the clutch there was a choice of dual-plate with a main drive plate of 279mm, while for the PTO it was 228mm. The alternative was a single-plate split-torque clutch giving a truly independent PTO. Earlier tractors had a six-speed forward and two reverse gearbox which with the addition of Multi-Power (at extra cost) doubled the number of speeds. Later models were fitted with MF's eight-speed forward and two reverse gearbox. All models had cast steel wheel centres front and rear. Tyre sizes varied according to model type. Power steering was standard across the range as were differential lock, front weight frame and

The US-built MF175.

The MF180 produced in North America was powered by a Perkins A4-236 diesel engine.

spring suspension seat. It is worth noting that the Low Profile model was equipped with shell-shaped rear fenders and headlights mounted externally either side of the radiator grille. Other models in the range had flat-topped rear fenders with headlights built into each forward facing end.

MF180

From North America in the early 1970s we have the MF180 Western, being the Standard model but also available in Tricycle, Wide Axle, and Row Crop. This range could be supplied with either the Perkins AD4-236 diesel engine giving 63bhp at the PTO or the Perkins AG-236 petrol engine unit giving the same output. Earlier models were fitted with a dual clutch together with the option of either a six-speed forward and two reverse or at extra cost the fitment of Multi-

The Italian-built MF184 4WD.

Power to double the number of ratios. At a later stage of production a single-plate split-torque clutch was fitted, with the PTO fully independent and operated by its own hydraulic clutch pack.

The tractors' hydraulic system gave Draft, Response, Position and Pressure Control, with the linkage having provision for both Category 1 and 2. On this range power-assisted steering had given way to full hydrostatic steering, which was a standard fitment along with differential lock and a spring suspension seat; the brakes were double dry discs.

MF184

This model was produced in Italy between 1980 and 1988 in both 2WD and 4WD versions. Both were powered by the Perkins A4-236 diesel engine with a quoted output of 68bhp DIN at 2200rpm. The clutch was the dual type giving independent PTO, with the facility to select PTO drive relative to engine speed or to ground speed. The earlier models up to about 1985 had as a standard fitment a second PTO mounted above the normal shaft, driven direct off the engine via a dog clutch, intended for belt pulley work. The gearbox was of the constant-mesh type with twelve forward and four reverse speeds and synchromesh between the forward ratios. Brakes were double dry discs. Interestingly, the hydraulic lift pump was a gear type mounted directly on the engine; the lift arms were Category 2. Tyre sizes were front 7.50 x 16, rear 16.9/14 x 30. The 4WD model had the same specification but the front tyre size was 12.2/10 x 24 and a mechanical linkage engaged the drive to the front axle; on this model hydrostatic steering was an optional extra.

A later development was the MF184F, again offered in two- and four-wheel drive. The main difference was in the styling of the sheet metalwork, which gave it a striking and individual look quite unlike any other MF tractors of the time. An interesting feature was the stylish and functional design of the rear mudguards: the inner section was made of steel while the outer part was formed in a hard rubbery plastic, as found on many modern tractors today. The mechanical specification was virtually the same as the MF184 but the brakes on the MF184F were three-plate discs, oil-cooled. The hydraulic linkage was Category 1 and 2. Hydrostatic steering was standard on all these tractors. On the 4WD tractors the differential in the front axle locked automatically but had mechanical engagement.

Towards the end of production, in 1988, MF of Italy introduced the MF184S. The general specification was more or less similar to the MF184F but there was a two-speed PTO running at 540 or 1000rpm. The linkage was Category 1 or 2. On the 4WD model an automatic engagement of the front axle could be installed as an extra. Also listed as extras were a creeper gearbox, front mudguards and extensions to the rear fenders.

MF185

The MF185 and its variants were introduced in 1972 and were produced in Mexico, the UK and France. The Coventry-built model is the subject of Chapter 9. Just to note some variations, the Mexican-built tractors had dry disc brakes whereas those produced in the UK and France were equipped with five-plate oil-cooled brakes. Power steering was optional as it was on those models built in the UK and France. Most had pressed steel wheels front and rear. Coventry-built MF185s for export to Sweden were badged MF185S or in turbocharged form MF185S Turbo.

The engine used to power the MF185 was the Perkins AD4-248, which in normally aspirated form had a quoted output of 71bhp DIN at 2000rpm, but when fitted with a turbocharger produced 84bhp DIN at 2000rpm. The transmission on the Mexican-built tractors was via a dual clutch to a six-speed forward and two reverse gearbox. MF185s manufactured in France and the UK offered users the choice of either a dual clutch mated to an eight-speed forward and two reverse gearbox, thus giving live PTO that could be driven relative to engine speed or to ground speed, or a single dry plate split-torque clutch giving a truly independent PTO. This was mated to an MF six-speed forward and two reverse gearbox compounded by the addition of Multi-Power. Tyre sizes on offer were generally 7.50 x 16 front, and 18.4/15 x 30, 16.9/14 x 34 or 13.6/12 x 38 rear. Weights quoted varied depending on final specification but around 2800kg would be the average. Most models were fitted with pressed steel wheels but front wheel weights and cast rear wheel centres could be ordered.

MF188

This was produced in France and the UK. The Coventry-produced MF188 is the subject of Chapter 8. Here we will consider briefly the variants produced at Beauvais. The 188, largest in the Super-Spec Range, benefited, like those produced in the UK, from the extra 6 inches (150mm) of wheelbase gained by fitting a spacer between the rear of the gearbox and the front end of the rear transmission case. Some of the French-produced tractors in this group were exported to Italy under the Landini dealership; they had slightly different styling to the front radiator grille in that the headlamps were bracket-mounted either side of the radiator side panels and not built into the grille mesh. All models in the group were powered by the Perkins A4-248 engine rated at 71bhp DIN at 2000rpm. Either a dual or split-torque single-plate clutch could be specified, coupled to an eight-speed (or twelve-speed with Multi-Power) gearbox. All models had cast wheel centres, the rears being of the PAVT type. Tyre sizes were front 7.50 x 16, rear 18.4/15 x 30, 16.9/14 x 34 or 13.6/12 x 38. The weight was 2858kg. Finally it should be noted that Beauvais did produce a MF188 with four-wheel drive. The drive to the front axle was engaged by a hydraulic clutch pack. Needless to say, on the 4WD version power steering was a standard fitment, unlike the MF188s where it was an optional extra. The front axle hubs contained epicyclic reduction gearing.

MF194

The final model in this overview of MF wheeled tractors in the 100 Series is the MF194 and its variants produced in Italy from 1980-1988 as the MF194S. These were powered by the Perkins A4-248 diesel engine rated at 73bhp DIN at 2100rpm. A dual clutch gave an independent PTO, driven relative to engine speed or ground speed. The earlier models featured a second PTO shaft extending above the normal PTO with a direct drive from the engine, intended for belt pulley work. This facility was deleted on the later MF194S models. The 194s were available in two-wheel or four-wheel drive throughout their production. Later models were fitted with three-plate oil-cooled disc brakes, hydraulically operated, while the earlier MF194 had double dry disc brakes. The hydraulic system on all the models in this group was fitted with an engine-mounted gear pump, with Category 2 linkage on earlier models, and 1 and 2 on the MF194S. These later models could also be supplied with a creeper gearbox giving twenty forward and eight reverse ratios.

Italian-built MF Crawlers

A range of MF crawlers was manufactured by MF in Italy from 1970 through to the 1980s.

MF101C

Built from 1970, this was powered by a two-cylinder diesel engine with a quoted maximum output of 25bhp at 2200rpm. It had a six-speed forward and two reverse gearbox, and two PTOs, one at engine speed and one controlled by its own clutch. Track steering was activated by lever control of multi-plate clutches. Basic equipment included a full lighting set, hours counter, and trailer braking unit. As extras one could order a dual clutch, a three-point hydraulic linkage, a front towing hitch and a downswept exhaust as an alternative to the normal vertical pipe. Weight without linkage was 1810kg.

MF124C

Also from 1970, this was powered by the Perkins AD-152 engine with fully independent PTO and a six-speed forward and two reverse gearbox. Hydraulic linkage was an extra but featured both Draft and Position Control. Unlike the MF101C, which did not have a top track idler, the MF124C did, one to each track, giving support and reducing track wear. This model weighed 2330kg.

MF134C

The MF134C crawler became available in 1975 and was produced in Standard, Narrow and Vineyard variants. Again these were all powered by the ubiquitous Perkins AD3-152 engine with a quoted output of 44.5bhp DIN. The clutch was a single dry plate split-torque type of 254mm diameter, giving a truly independent PTO. Drive was taken through an MF eight-speed forward and two reverse gearbox. Three-point hydraulic linkage was standard, offering Draft and Position Control. Track steering was controlled by two levers operating dry multi-plate clutches. The Standard width model stood 1.18 metres, while the Narrow and Vineyard stood at 1.17 metres. Another model of this period was the MF144C, again powered by the Perkins AD3-152 engine. The clutch was a single-plate split-torque type giving a fully independent PTO.

The gearbox was the MF six forward and two reverse unit, the steering by multi-plate clutches was lever controlled, and the tracks were 310mm wide, supported by four bottom rollers and one top idler per side. Hydraulic linkage with Draft and Position Control was standard. These models were only manufactured in one width, 1.410mm overall, and weight was 2879kg.

MF154C

This crawler was produced in Standard, Narrow and Wide variants, all powered by the Perkins AD3-152 engine, with a single-plate split-torque clutch operated by a hand-lever, and an eight-speed forward and two reverse gearbox. Steering was again by multi-plate dry clutches. Hydraulic linkage was not fitted, nor was it listed as optional. The width of the Standard model was 1310mm, with a quoted weight of 2930kg.

An Italian-built MF134C about to go disc harrowing powered by a Perkins A3-152 engine.

MF 164C

The Perkins AD4-203 engine was fitted to this variant, as used in the earlier Coventry-built MF165, with a quoted power output of 60bhp at 2000rpm. The hand-lever controlled clutch was of 310mm diameter, and there was live PTO, engine speed or ground speed related. Hydraulic three-point linkage was standard, with Draft and Position Control. Track steering was by multi-plate clutches, controlled by levers. The tracks were equipped with five bottom rollers and one top idler per side. The tractor weighed in at 3107kg, with an overall width of 1460mm. As on all the Italian-built crawlers there was a full lighting set.

MF174C

Introduced in 1975, the MF174C was fitted with the Perkins A4-212 engine producing 61bhp DIN at 2200rpm. It was available in Standard, Narrow and Wide versions, widths being 1410mm, 1310mm and 1660mm respectively. The respective weights were 3230kg, 3220kg and 3340kg. The single-plate split-torque clutch, lever-controlled, delivered drive to an eight-speed forward and two reverse gearbox. Multi-plate clutches controlled the track steering, with each track having five bottom rollers and one top idler.

MF184C

By 1980 MF in Italy were manufacturing the MF184C crawler, powered by a Perkins A4-248 diesel engine with a rated output of 73bhp DIN at 2100rpm, the same as fitted to the Banner Lane-produced MF185. A split-torque clutch gave a fully independent PTO. Hydraulic linkage was standard, offering the usual Draft and Position Control. As optional equipment a two-speed PTO could be factory installed.

MF1114C

A larger machine from 1980, this was powered by the famous Perkins A6-354 direct-injection engine with a quoted power output of 100hp DIN at 2250rpm. The clutch was a four-plate wet-disc type of 355mm diameter, which passed power to a six-speed constant-mesh gearbox coupled to a hydraulic shuttle. The PTO was live, related to engine speed. Track steering was by hand-lever operated servo-assisted multi-plate wet-disc clutches. The standard track shoes or grousers were 510mm wide but 610mm could be specified. These crawlers came as standard with only auxiliary hydraulics, which supplied a four-spool valve block with a flow rate of 90litre/min. Three-point linkage was an extra-cost fitment as was a two-speed PTO giving 540 and 1000rpm. The overall width of the standard models in this range was 2030mm, and weight 7250kg.

MF1124C

Launched in 1980, this was the largest crawler produced in that era in Italy. Its power unit was the direct-injection turbocharged Perkins AD6-354.4, producing a quoted 124bhp DIN at 2250rpm. In general terms the specification follows that of the previously outlined model though upgraded in some areas to cope with the extra power. One of the main differences was in the track specification: six bottom rollers and two top idlers per track were fitted. Standard track shoe or grouser width was 510mm but 610mm shoes could be fitted at extra cost. The auxiliary hydraulic system was similar to that just outlined but with the flow rate increased to 153 litre/min. The overall width of the MF1124C was 2040mm on standard shoes, with a weight of 9700kg. As accessories a three-point linkage could be installed, as could a dozer blade, two speed PTO and road pads.

MF1080 with a Perkins A4-318 direct injection diesel engine.

Other Models

To consider in any detail all the other models of wheeled MF tractors that were produced in the UK, France, Italy, North America, Argentina, Germany and Mexico in rigid or articulated form is beyond the scope of this book, but there are a couple of models I would like to mention, including the articulated tractors.

The MF1080, made in France 1968-72, was powered by a Perkins A4-318 diesel engine with a quoted output of 88bhp DIN at 2000rpm. The clutch was a single-plate split-torque type, the gearbox had twelve speeds with Multi-Power, and the steering was fully hydrostatic with mechanical back-up. The hydraulic linkage was a heavy-duty version of MF's normal design with Category 2 hinged and telescopic ball ends, but Category 3 could be specified. A two-speed PTO was standard. The Beauvais factory also produced a 4WD version of this model, front tyre size being 12.4/11 x 28, rear 16.9/14 x 38 on the 2WD model.

Turning to the articulated tractors that were produced at MF's industrial facility at Barton Dock Road, Manchester, it is appropriate to mention that the articulated principle as applied to tractors was first developed by the Mathew Brothers, otherwise known as Matbro. They developed this system for their articulated four-wheeled loading shovels. Caterpillar copied the Matbro design, which of course they had patented, and this infringement provoked a court action by the Mathew brothers, who rightly gained a massive compensation payment from Caterpillar – a close parallel to Harry Ferguson taking on Henry Ford II for stealing his designs to incorporate into Ford tractors.

The MF1200 of 1972 was MF's first foray into an articulated design, and was powered by the Perkins A6-354 direct-injection engine with a quoted maximum output of 105bhp DIN at 2400rpm. These tractors used many components common to the larger MF100 Series manufactured at Banner Lane. The clutch was a single split-torque type giving true independent PTO.

The standard gearbox was the MF eight forward and two reverse speed type, but as an alternative the tractor could be built with twelve forward and four reverse ratios incorporating Multi-Power. The brakes, on all four wheels, were of the dry disc type, hydraulically operated. The PTO ran at 1000rpm and steering was fully hydrostatic. The hydraulic system was the normal heavy-duty three-point linkage featuring Draft, Position, Pressure and Response Control. An external linkage assistor ram gave the Category 2 linkage arms with telescopic ends the ability to lift 3702kg. Tyre sizes all round were either 13.6/12 x 28 or 18.9/14 x 34.

The MF1200 was equipped with a well insulated safety cab that included tinted glazing, a fresh-air blower and a heater – even air conditioning became an option. These tractors stood 2.9m high and weighed in at 5156kg.

By July 1979 the more powerful MF1250 had been introduced, again powered by the Perkins A6-354.4 (the suffix 4 denoting Mk IV in the development of that engine). This particular engine had a quoted output of 112bhp DIN at 2500rpm. The clutch was the single-plate split-torque type of 330mm diameter. The PTO ran at 1000pm at 2216 engine rpm with its own brake incorporated in its drive clutch pack. The gearbox had twelve forward and four reverse ratios incorporating Multi-Power. Three tyre sizes were offered: 13.6/12 x 38 eight-ply radial, 16.9/14 x 34 eight=ply radial or 18.4/15 x 30 six-ply crossply. The height of this model over the cab was 5.18m, weight 6202kg. The hydraulic system was basically the same as on the MF1200 but the lifting capacity on the lower links was increased to 4375kg. Tyre sizes were the same as on the MF1200. The brakes, on all four wheels, were five-plate oil-immersed, hydraulically operated.

The Detroit factory produced two larger articulated models, the MF1800 and the MF1805, both equipped with Caterpillar V8 engines.

I do hope that this "brief overview" gives readers an insight into the vast range of tractors that MF manufactured in the era covered by this book, as well as a sense of the international scale of their design, development, production and marketing endeavours. One can only speculate on the numbers of people they employed directly, but also indirectly with subcontractors, to achieve the company's tremendous standing in the agricultural and construction market.

This is the front cover to a British 1976 MF1100 sales brochure. Although these models were built in North America they were modified in the UK to meet the UK legal requirements, particularly in the area of lighting: note the front mounted head lights and blanking plates on the rear mudguards, now with only a side light on each wing.

The MF1150 used the massive 146hp 8.3-litre Perkins AV8-510.

Chapter 2

The MF130 – built in France

The MF130, the smallest of the Red Giant range of MF tractors, was launched alongside the others at the 1964 Smithfield Show. It was the one model that had French ancestry, and in many ways it was quite different from the more powerful models built at Coventry, but its styling was in line with the rest of the range. It was in reality a developed version of the French built MF825, which had been popular on small French farms, but many improvements were incorporated into the Beauvais-built MF130.

The engine was the Perkins four-cylinder diesel AD4-107, the same as used in the MF825 but with its governor set to give a higher maximum speed of 2250rpm, at which it produced 30bhp, 5bhp more than the earlier model. This engine was an indirect-injection type with wet cylinder liners. The bore was 3.125in (79.38mm), stroke 3.5in (88.9mm), compression ratio 22.5:1, and it was fitted with a CAV distributor-type injection pump with built-in mechanical governor. Cold starting was aided by a CAV Thermostart device fitted to the induction

An MF publicity photograph of an MF130 towing an MF7-71 3-ton tipping trailer.

manifold. It had a cast steel sump so that the unit-construction principle that is normal practice in tractor design could be employed. The earlier MF825 model had a pressed steel automotive-type sump, and an additional fabricated steel frame had to be used to join the front axle pivot to the transmission case.

The clutch could be of two types. The Deluxe tractor had a dual clutch, giving live PTO, with a main clutch of 11in diameter and a 9in PTO clutch. The Economy model had a single 11in clutch. The gearbox had eight speeds forward and two reverse, similar to the previous MF825. An unusual feature at the time was that synchromesh was provided between third and fourth gears and between seventh and eighth, controlled by the main gear lever and the high-low selector lever, which had to be set in neutral before the starter could be operated.

The hydraulic system, although incorporating two-way Draft, Position and Response Control and an oil supply to three hydraulic take-off points, was of a very different layout from that used on other models in the 100 Series range. For a start all the components of the hydraulic system are housed within the lift top cover, which can be removed as a unit once the lift links are disconnected and the 14 bolts holding the top cover to the transmission case are taken out. Incorporated into this design is a mechanical lift lock to hold the lift arms in the fully raised position, thereby taking the strain off the system when transporting heavy implements at speed over rough terrain. Also incorporated within the hydraulic circuit is a filter on the suction side of the pump and a full-flow filter on the output side. A thermostatically controlled oil cooler is located in front of the radiator, the idea being that the common transmission and hydraulic oil run at optimum temperature. The water radiator and oil cooler can easily be cleaned by opening the hinged radiator grille.

Below is a description of this rather unusual MF hydraulic system. The main single control lever, set within a quadrant, operates two ways, giving Draft, Position, and external services control as per the accompanying diagram (right). In addition to this main control lever there is a lever-type control on the left-hand side of the driver's seat that can be set to vary the speed of drop of an implement, while on the right-hand side of the lift cover there is a small lever that gives a degree of Response control, i.e. sensitivity to the draft control mode. Response Control varies the reaction time, i.e. heavy plough, slow; light cultivator, fast. It should be mentioned here that the two-way Draft Control on this model is achieved not by the use of compression springing but by the neat expedient of

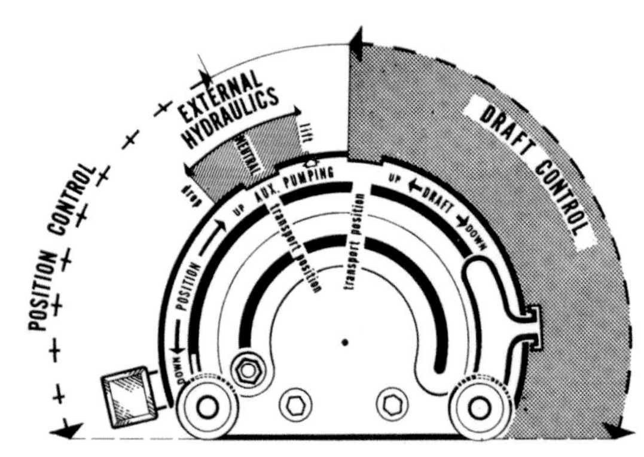

The centrally mounted hydraulic control quadrant on an MF130 Vineyard tractor. Note the Bowden cables, not exactly the most convenient position! The Coventry-built MF135 Vineyard had a much better arrangement.

The Lift Lock incorporated into the top cover design to take the strain off the hydraulics when in transport mode.

The Coldridge Collection's MF130. Early models had red wheel centres.

The offside view of the indirect-injection Perkins AD4-107 engine.

a double-acting leaf spring.

Turning now to the final drive components, we find the power take-off has two modes of operation, selected by a lever located on the right-hand side of the transmission housing. The PTO is driven either proportional to engine speed, so that 1890rpm engine speed gives the standard PTO speed of 540rpm, or proportional to ground speed, so that on 10.00 x 28 rear tyres the PTO rotates one revolution per 18 inches of travel. I can find no reference to a mid-mounted PTO, an option which was available on the earlier MF825. To me the underside of the MF130 is identical to that of the MF825. An MF publication relating to the MF825 issued to dealers in 1961 states, "Underbelly PTO: the forward end of the main PTO shaft is splined to carry the drive gear for the underbelly". The PTO shaft is carried in a detachable housing bolted below the main transmission casing; at an engine speed of 2000rpm the PTO ran at 1057rpm.

The brakes on the MF130 were most unusual: they were inboard within the trumpet housing, close to the differential. Attached to the halfshafts is what can only be described as a heavy vee pulley into which is pushed, by operation of the brake pedal, a small chunk of tapered friction material. Although these brakes were prone to rapid wear, at least the friction material could be replaced easily by undoing the three bolts that retained the actuating mechanism and brake pads to the trumpet housing. The brakes could be

MF130 dash retaining its French-made instruments.

New styling for the 100 Series brought the MF825 into line with new range.

A view of a dismantled epicyclic rear hub, showing clearly the three planet gear wheels giving a reduction of 8:1.

used independently or latched together for road use; a parking brake lever locked the brake on positively. In order to meet Road Traffic Act requirements for two independent braking systems to be fitted, the industrial MF2130 imported to the UK was equipped with additional 14in (355mm) drum brakes fitted to the rear wheel hubs and engaged by the parking brake lever. On these industrial models the sheet metalwork was painted in MF Highway Yellow.

These small MF tractors were fitted with epicyclic reduction hubs at the rear wheels. Substantially built, with three planet gears giving a reduction of 8.1:1, they enabled a lighter differential to be fitted which, by the way, incorporated a locking facility. This was

An MF130 pulling a MF793 plough at Stoneleigh.

standard on all models.

The front axle is rather like a smaller version of that fitted to the MF165, being made in three pieces. The centre section that pivots is of hollow rectangular section, while the two outer sections that carry the swivel pins are made of solid steel. The axle can be adjusted in (or out) in 4-inch (101mm) increments from 48in (1220mm) to 72in (1830mm) on the Standard and High Clearance tractors. Adjustment is of course more limited on the Narrow, with front track setting from 42 to 63 inches (1.01 to 1.62m), and on the Vineyard model, which had a front track setting of 29.5 to 45 inches (0.75-1.13m). The steering box is a worm and sector Gemmer type with a double bearing to compensate for wear. Numerous tapped holes were incorporated into the side and underside of the transmission housing, as well as the fixing points provided on the cast steel front axle pivot part. These holes were carefully arranged so that a number of implements or accessories could be fitted simultaneously.

To round off this chapter let us consider some features that made the driver's life a little more pleasant. If he was lucky enough to be using the Deluxe model he would have the convenience of live PTO, a cushion seat, a screw-adjustable top link and road lights, but the ploughing lamp was an extra! His instrument panel would have an oil pressure gauge, a temperature gauge, an ammeter, a tractormeter, a hand throttle and an engine stop control, as well as a lights switch that incorporated the horn button. Shell-type rear fenders reached down to the foot boards. The bonnet top was hinged on the right-hand side so that it could be raised well clear of the 10-gallon (45.5-litre) fuel tank. The radiator grille was also hinged on the right and could be swung open to clean the oil cooler and radiator fins as well as providing good access to the oil bath air filter. Conversely his less fortunate brother driving the Economy model had to make do with a steel pan seat, a single clutch, no lights or horn and a two-piece top link. Not a great number of these Economy tractors were sold in the UK.

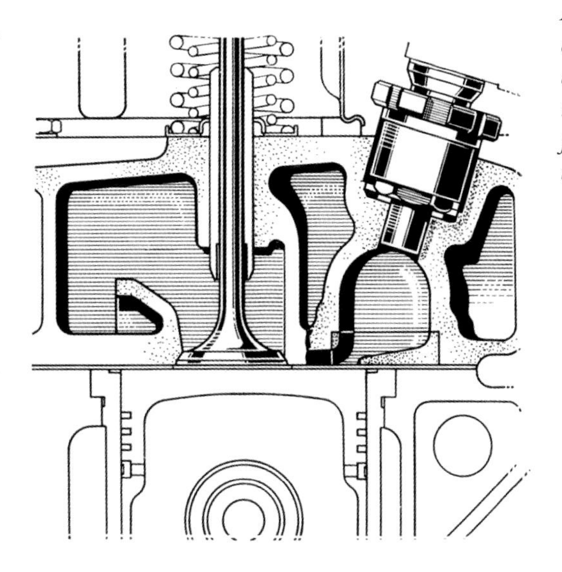

A cross-section drawing of the Perkin H-type combustion chamber; being indirect it permits the use of flat topped pistons to ensure even heat distribution.

The engine gained a somewhat dubious reputation in the UK, often suffering failure due to heavy sludge formation in the sump, a problem that the French denied. The problem seems to have been related to the fact that the vast majority of MF130 and MF2130 sales in the UK were for parks and gardens and golf course work, with the minimal loadings that these usually demanded. In France, to be fair, they were probably worked somewhat harder, which was beneficial. To get round this, the engine oil change period was reduced from the standard 250 hours to 100 hours, which helped.

It is worth comparing relative prices – in December 1964 the MF130 Deluxe diesel was £660 and the MF135 Deluxe £740, so for an extra £80 one could buy what in my opinion was a much more robust, reliable and powerful tractor.

This photograph illustrates the external layout of the hydraulic lift assembly.

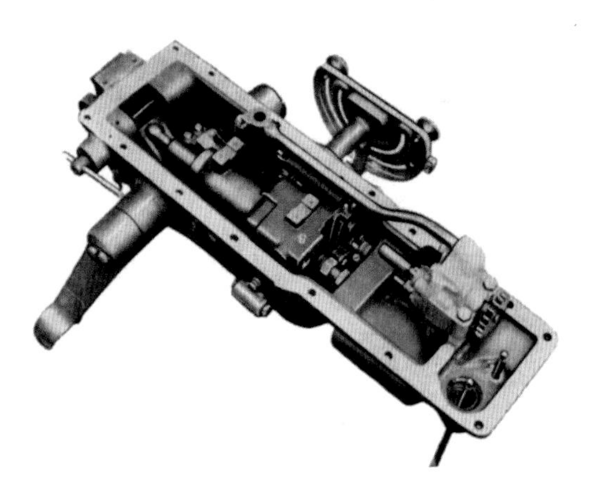

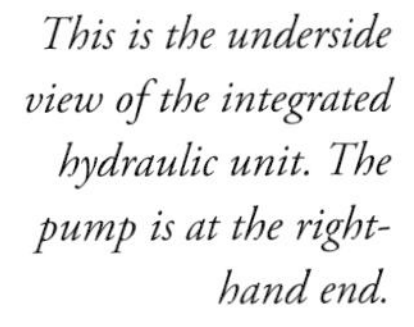

This is the underside view of the integrated hydraulic unit. The pump is at the right-hand end.

Chapter 3

The Coventry-built MF135

Being a biased Ferguson enthusiast I will take the risk of stating here that the MF135 is possibly the most sought-after and coveted of the early MF range of tractors. If this be so it is not without justification, for surely the 135 combines all the clever pioneering work of Harry Ferguson and his team of engineers, as well as the single-minded determination of Herman Klemm and his engineers in Detroit, who further developed the hydraulic system and the power output of the engine to design and build what eventually became the TO35 and FE35 Ferguson tractors. These models in their turn were further refined over the eight-and-a-half years of their production, finally forming the basis of the MF135 launched in London at the 1964 Smithfield Show along with the other members of The Red Giant range. All these tractors had a strong, stylish, purposeful aesthetic to the sheet metalwork of the bonnet and radiator grille – a kind of "We mean business" as a visual statement.

The Coventry-built models shared a lot of common

The dispatch parking area at Banner Lane – just try counting the number of tractors in the photograph!

Front three-quarter offside view of Jonathan Lewis' original and early MF135.

features and evolutionary changes over their long production run from 1964 to 1979. Let us first consider the features of the engine, noting along the way the subtle changes that were implemented as time went on until the end of production. The engine generally installed was the Perkins AD3.152 three-cylinder direct-injection diesel, but a few MF135s produced at Coventry were powered by the Standard Motor Company's four-cylinder ohv 87mm-bore petrol engine, mainly to meet export requirements, particularly Denmark and New Zealand, although a few found buyers in the UK.

The Perkins AD3-152 had a bore of 3.6in (91.44mm) with a stroke of 5in (127mm) giving a capacity of 152cu.in (2.5 litres), hence its designation: AD for diesel, 3 cylinders, total capacity 152cu.in. The compression ratio was 18.5:1, firing order 1,2,3, and maximum output 45.5bhp at 2250rpm. The Perkins AD3-152 had proved itself to be a robust and reliable unit, having been successfully chosen for the later MF35 diesel models, with something of the order of 342,400 being installed in this application alone. We can add to that figure all the engines that were used by the many manufacturers of machines in the realms of agriculture, industry, marine and construction. Further afield, there were numerous licence agreements set up with engineering companies in India, Yugoslavia, Turkey, Japan, Argentina, Bulgaria, Mexico, South Korea, Peru, Pakistan, Poland and Iran, not to mention others that were set up after 1979 and therefore outside the scope of this book.

So what made this engine so outstanding? Well, to quote an MF publication, "Having a direct injection system, all the combustion chamber clearance volume is

Massey Ferguson's demonstration of their pressure control system at the Royal Show in 1966.

Front and rear views of this June 1966 MF135. The rear view complete with Devon registration plate.

Proof that plastic grille badge and bar were fitted to some early MF135s.

The front weight frame with genuine MF draw pin in place.

Early type Butler head lamps with aluminium surrounds.

The dash panel with centrally-positioned tractormeter, as fitted to early models.

The early type cast aluminium grab handle incorporating a side light.

A clear shot of the response control quadrant.

Early MF135 tractors were fitted with bonnet flap fasteners identical to those fitted on later MF35s. Eventually a slightly longer type became the norm.

A rare example of the sprung hinge battery cover, again only fitted to early tractors but often removed to simplify battery removal and replacement!

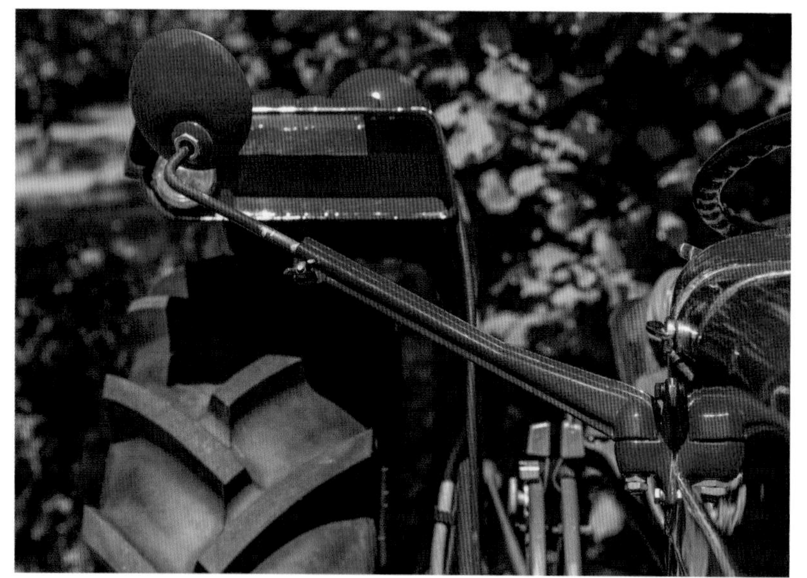

The original early type rear view mirror made by Wingard.

A close-up of the tractor's swept back front axle – similar to that of the TE20.

The Ferguson System decal as positioned on early MF135s. Later models had it included on the silver MF strip.

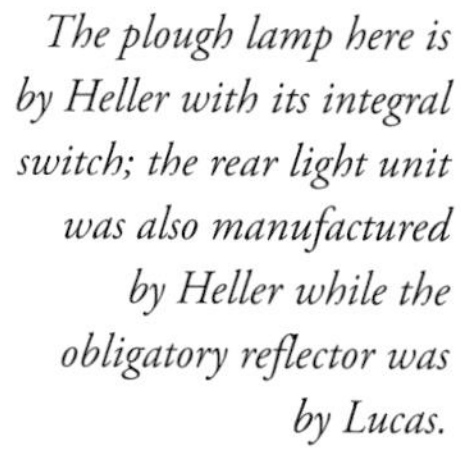

The plough lamp here is by Heller with its integral switch; the rear light unit was also manufactured by Heller while the obligatory reflector was by Lucas.

Note the red wheel centres of this early model which is fitted with inside rear wheel weights.

The early type of two-piece PTO guard, with the sheet metal surround supported at the top by a cast steel bracket.

The twin fuel filters; note the primary one with a transparent bowl to collect and show sediment.

contained within the depressions in the piston crown. This design of combustion chamber has the lowest ratio of surface to volume and the heat losses are relatively small. For this, the thermal efficiency for a given compression ratio (18.5:1) is very high. This means low specific fuel consumption and the best possible starting from cold."

The architecture of this engine featured dry liners to each bore. The valves were of the overhead type arranged to give a cross-flow layout with the exhaust manifold to the left-hand side when viewed from the tractor seat. The air induction from the right was fed via an oil bath type air cleaner up to model number 162200; after that point, with the introduction of the updated MF135s in 1971, starting at serial number 400001, the oil bath air filter was replaced with a dry element type with replaceable cartridge and a service indicator just below the dash panel.

The crankshaft of the AD3-152 engine is of forged steel running on four journals fitted with replaceable shells, as are the big and small ends. The camshaft is driven via an idler gear; a second sprocket takes drive from the idler to operate the fuel injection pump.

The lubrication system of the Perkins AD3-152 was conventional, having a wet sump with an oil capacity of 10.5 pints (5.97 litres). Three filters were incorporated within this engine: a coarse mesh strainer at the base of the oil filler tube; a wire mesh strainer in the bottom of the sump which could be removed for servicing; and the main full-flow externally-mounted filter with

The early two-piece number plate bracket assembly.

A near-side view of the Perkins A3-152 engine with the Multi-Power pipes to the oil cooler and filter assembly.

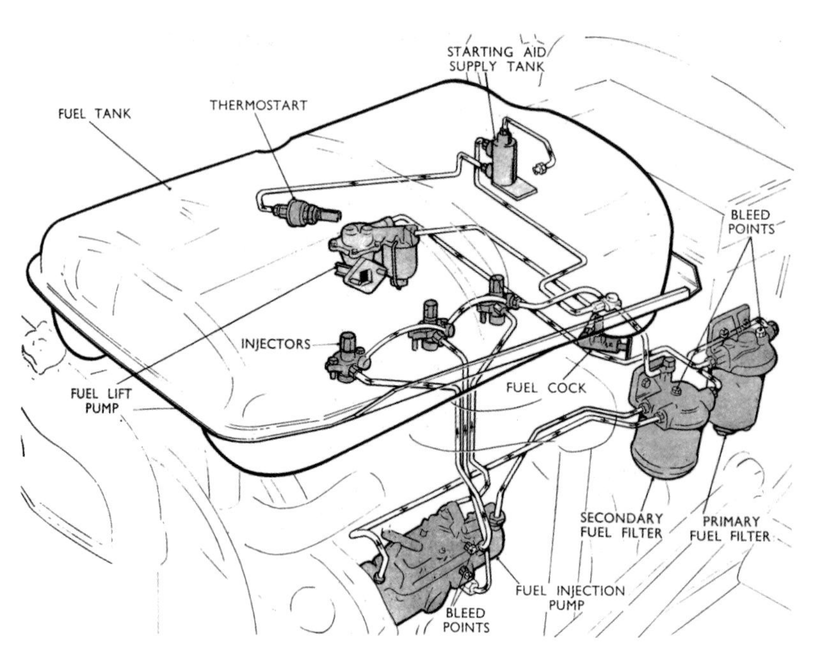

This diagram of the fuel system is taken from the MF35 Workshop Manual but it is very similar to that of the MF135.

a replaceable cartridge, mounted on the left-hand side of the crankcase. Thus engine oil is filtered under pressure. Should the filter element become totally choked, pressure would build up within the inlet port of the filter; so as a safety precaution a spring-loaded ball bypass valve was fitted which allowed unfiltered oil to continue to circulate.

The oil pump is fixed to the front main bearing cap within the sump and takes its drive via an idler gear from the crankshaft's spur gear. The pump's impeller is a four-lobe rotor which meshes with a five-lobe rotor within the pump body which is free to rotate within the pump body. As this rotor rotates the spaces formed between the lobes increase and decrease in volume, causing the oil to be taken from the suction side to the pressure side of the pump. The oil pump is fitted with a pressure relief valve that opens at 50-65 psi; it has a delivery capacity of 5.35 gal/min (24.3 lit/min).

As far as the cooling system is concerned, it was a conventional pressurised arrangement with a belt-driven water pump with a two-bladed fan attached. In the upper part of the cooling circuit was a bellows-type thermostat to assist with rapid warm-up of the engine; this started opening at 175°F (79.5°C) and was fully open at 200°F (93.5°C). An engine temperature gauge was a standard fitment.

The fuel-injection pump was a CAV distributor type with a mechanical governor. The fuel system comprised of a fuel tank of 8½ gallons (38.61 litres) on earlier models, being increased to 10 gallons (45.41 litres) in 1971 when updated models were introduced. The fuel lift pump, by AC Delco, takes diesel to a pair of replaceable paper element filters arranged in series, then on to the injector pump. Excess fuel from the pump and injectors was returned to a small starting aid tank which had a return to the main tank; this small tank served as a feed for the Thermostart, which was an aid to cold starting. This device was screwed into the aluminium induction manifold at the air intake

These photos show the 1965 MF135 from the Coldridge collection. It has the Standard petrol engine, as used in the Petrol/TVO MF35.

end. In general it would only be used when ambient temperature dropped below 32°F (0°C).

The workings of this device, made by CAV, are best understood by making reference to a numbered cross-section diagram. It is operated by the starting switch on the dash panel. When turned anti-clockwise to the first setting heat (H) position for about 10 seconds; a current is applied to the unit that contains two heating elements and a solenoid to control the flow of diesel taken from the small tank.

Basically the Thermostart consists of a core (3 in the illustration), a solenoid (6), spring-loaded plunger (4) with a special rubber insert (5) which butts against a valve seat (7). A coil carrier (8) has two heating elements (9) and (10) and a circular shield having large perforations (11) on one side and small perforations (13) on the other and a small flange (12) running along its other surface.

In operation, fuel from the small tank fills the pipe adapter (1), filter (2) and hollow plunger and the grooves on its surface (4). With the first turn of the starter key anti-clockwise to the H position the solenoid (6) and heater coils (9, 10) are energized. Magnetism induced in the plunger (4) and adapter (1) by the solenoid draws the plunger and rubber insert off the valve seat (7). Fuel then flows at a controlled rate along and around the coil (9) which vaporises it. The coil (10) reaches the ignition temperature of the fuel vapour. As soon as the engine is cranked by turning the key to it full extent anti-clockwise i.e. H to S (Heat to Start); fresh air drawn into the inlet manifold enters the circular shield through the small perforations (13) and mixes with the vaporised fuel inside it. The resulting mixture is ignited by the coil (10) and so heats the air for combustion as it is drawn into the cylinders, promoting easier ignition of the injected fuel. The flange (12) running along the outer surface of the shield provides a shelter zone around the outlet holes (11) and protects the flame from the incoming air stream.

The 12 volt electrics were wired negative earth and tractors were fitted as standard with a 17-plate 96amp/

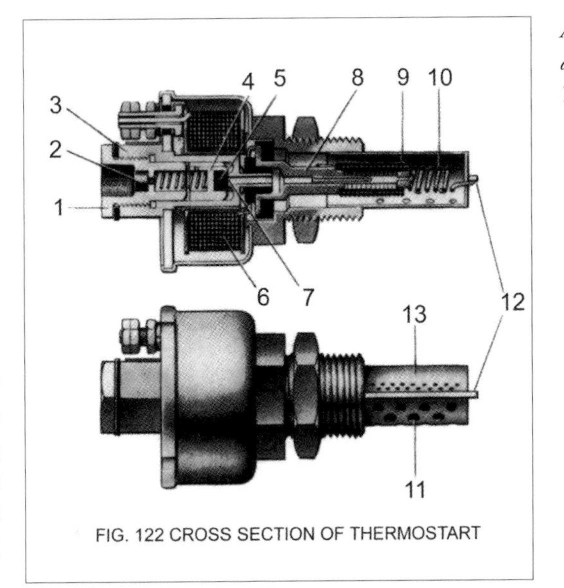

FIG. 122 CROSS SECTION OF THERMOSTART

A numbered cutaway drawing of the CAV Thermostart.

The swinging draw bar was a standard feature.

The nearside view of the Standard petrol engine.

Lucas distributor and coil. Note the take-off to drive the Tractormeter.

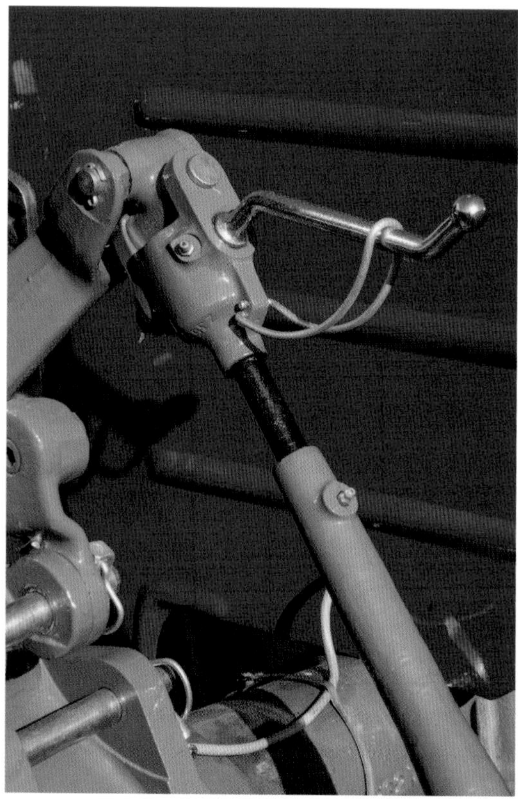

Spring-loaded loop locks the levelling box handle in position.

The Standard motor was fuelled by a Zenith 28G-2 updraught carburettor.

The ignition light to the right of the steering column was only fitted to petrol models. This tractor has no fuel gauge.

hour battery, but for cold climates a 17-plate 125amp/hour battery was installed. Normally fitted was a Lucas M45 pre-engaged starter motor activated by a key switch on the dash panel; the starter would only function with the high/low range gear lever in neutral. A CAV CA45 starter could be specified for cold climates. The dynamo was normally a 12-volt Lucas C40A with its output controlled by a Lucas RB108 regulator. An ignition warning light was only fitted to petrol models, which were equipped with a Zenith 28G-2 updraught carburettor with the choke control below the dash, and both diesel and petrol models had dashboard-mounted fuel, temperature and oil gauges, together with an ammeter.

Turning now to the transmission, the clutch on diesel models could be of two types. The single type was an Auburn of 11in (279mm) diameter, coil spring operated. The alternative, more generally fitted, was the Auburn dual type giving live PTO; the main clutch was of 11in (279mm) diameter, coil sprung; whilst the PTO clutch was a 9in (229mm) Belleville spring, operated and controlled by the clutch pedal. A Belleville spring is a ring of flat spring steel formed to have a slightly conical shape. In the case of the 135 fitted with a petrol engine, again there were two options: a single clutch of 9in (229mm) diameter or the Auburn dual type to the same dimensions as fitted to the diesel models.

The gearboxes of all models were the same, with three forward ratios and one reverse, compounded by a secondary planetary gear unit giving a reduction of 4:1 and thus six forward and two reverse speeds.

Multi-Power (a form of overdrive) could be ordered but only on dual-clutch models. Multi-Power was unique to MF when introduced in 1962. It gave the driver the facility to change the ratio of a selected gear on the move and under load, up or down, but not

Lucas C40A dynamo was controlled by a Lucas RB108 regulator.

Robert Perry's later 1971 diesel-engined MF135.

PTO-driven Newland saw bench.

between gears. The downside of this option was that engine braking was only effective when Multi-Power high was selected; in low there was no engine braking effect, hence the risk of the tractor and load running away downhill. The control lever was conveniently placed on the dash. MF branded Multi-Power "A Flick Change Transmission". Noting this potentially dangerous drawback the Ford people quickly retaliated with the quip, "A flick of the switch and you're in the ditch". However, to be fair, Ford did introduce their Dual Power system (an under-drive) in December 1973, eleven years after MF Multi Power, which was intrinsically safer in that engine braking was provided in both high and low ratios.

Differential lock was a standard fitment. The PTO was live on dual-clutch models and all had a selectable PTO that could be driven relative to engine speed or to ground speed. For export markets, particularly Germany and Austria, a mid-mounted PTO could be installed, ideal for mid-mounted mowers. The PTO dog clutch was behind the four-piston scotch yoke hydraulic pump, so on dual-clutch models it could be termed constant running.

The brakes on all models were mounted directly on the wheel hubs and were of the internal expanding type made by Girling, measuring 14in x 2in; they could be operated independently or latched together for road use. A parking brake lever was standard, linked directly to the foot-operated linkage. For specialist applications, such as working in paddy fields with deep water,

The Perkins A3-152 engine continued mainly unchanged but the rear crankshaft oil seal used in the later engines was a large, lip-type seal. Early engines used a rope seal.

Girling 9in x 6in sealed brakes were sometimes fitted. The other exception in the area of braking systems is confined to tractors for certain export territories and the industrial models, namely the earlier MF2135 and the later MF20 produced at Coventry. To comply with the UK Road Traffic Act, as road vehicles it was necessary to have two independent braking systems. This was achieved by having wider brake drums encasing two pairs of concentrically arranged brake shoes. The inner sets are 14in x 1½in, controlled by the handbrake lever, while the outer sets are 14in x 2in, controlled by the pedals, all being mechanically operated.

The steering box used throughout the production run of the MF135s was of the screw and recirculating ball nut type very similar to that used on the MF35s. The layout originated in the design offices of Massey Harris Ferguson in Detroit headed up by Herman Klemm. This type of steering box ensures that very little friction is generated, resulting in smoothness of operation. The "nut" that is wound up and down the thread cut in the steering shaft is attached to the left-hand rocker shaft; it has rack teeth on its upper side that engage with teeth cut in the lower side of the right-hand rocker shaft, which is above. To the splined ends of both rocker shafts protruding through the enclosed oil-filled steering box are attached the drop arms that in turn operate the drag links positioned either side of the engine. The drop arms are of different lengths, the left-hand slightly shorter than the right. The centres of the rocker shafts and the lengths of the drop arms are so arranged as to ensure correct steering geometry at all front axle width settings.

The hydraulic system incorporated the features from the MF35X, giving Draft, Position and external services control, but also available was a Response

Later MF135 dashboard with offset Tractormeter.

This motorised cutaway of an MF135 rear end was used at Lackham Agricultural College in Wiltshire to demonstrate the inner workings of the hydraulic system.

The Coldridge Collection's 1968 Vineyard MF135. This was Ernie Luxton's last restoration project for the collection.

Front and rear views showing the extreme narrowness of these models.

The reduction gearbox adds to the length of this example.

The Massey Ferguson reduction gearbox doubles the number of gear ratios available.

The unusual arrangement of the swivel arms, extending outward and over the front wheels. This necessitates the use of a special steering box where its drop links operate in the opposite direction to those on the regular MF135.

Cranked drag arm of the Vineyard MF135.

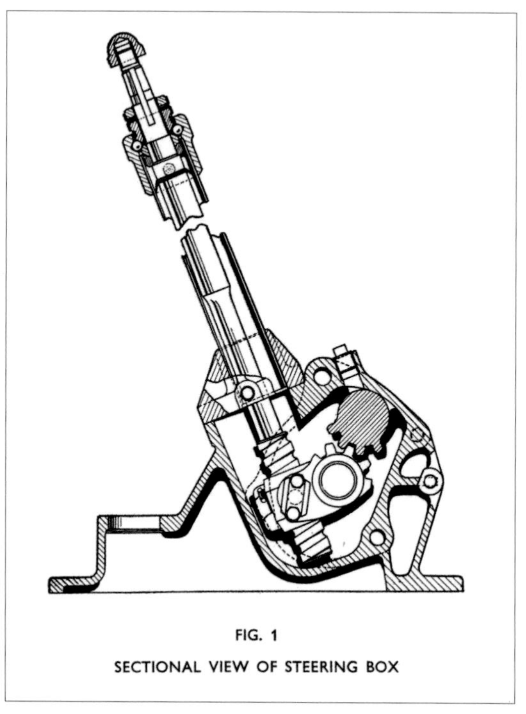

FIG. 1
SECTIONAL VIEW OF STEERING BOX

Showing the sets in the lower lift arms – note the added reinforcement and the extended handle of the levelling box.

control quadrant mounted on the right-hand side of the transmission housing. Another facility that could be incorporated within the hydraulics was known as Pressure Control, used in conjunction with an MF Pressure Control hitch to aid traction with trailed implements. Multi-Power tractors could be fitted with one or two types of auxiliary hydraulic pumps in addition to the main lift pump. Option one was a basic Multi-Power pump that just operated the clutch pack in that unit; this pump could not be used to power external services. Option two was a dual-element auxiliary pump which operates the clutch pack but also provides oil for the externally mounted spool valves. This circuit operates independently of the tractor's main lift pump. A combining valve could be installed to join together the outputs of these two pumps. The output of the linkage lift pump at maximum engine speed of 2250rpm is 3.3gal/min (15 litres/min); at the same speed the auxiliary pump's output is 7gal/min (31.3 litres/min), so combining the two outputs we have 10.3gal/min (46.7 litres/min) or a useful 14.4hp at 2000psi. As a comparison let us look at the output of the hydraulic system of an MF135 with a single clutch: the standard lift pump in this case is rated at 4gal/min (18.2 litres/min) at engine speed of 2250rpm, PTO speed of 810rpm. The standard three-point linkage was Category 1, but as mentioned in Chapter 11 a universal

Narrow MF135 from the Coldridge Collection with PAVT rear wheels, in original and unrestored condition.

The Narrow tractor with its track at the narrowest setting.

Note the slight cranking of the drag arm and the set at the clevis end of the radius arm.

A comparison of front views of the Vineyard, Narrow and Standard MF135.

Category 1 and 3 kit of parts was marketed by MF for those users who had that requirement.

The earlier model of MF135 had a front axle arrangement similar to that of the MF35s, i.e. a three-piece hot-forged set-up of swept-back design braced by a pair of tubular radius arms. This ensured that front axle track adjustments could be made without altering the steering geometry. This design was Ferguson patent number 541220 dated 8 July 1939. In 1971, when the updated MF135 was introduced, the front axle was radically changed. The Ferguson design gave way to a three-piece layout made up of a heavy central fabricated steel box-section pivoting central bearing, set within the cast steel front frame bolted directly to the engine. Into either end of the central box-section were fitted the outer axle assemblies, which again were of fabricated mild steel and held at the desired track setting by bolts passing through the centre section and the outer. This layout did not have the benefit of allowing front track adjustments without interfering with steering geometry. It did have the advantage of being more robust and eliminating the pair of radius arms, thus reducing production costs. It was possible to alter the front wheel track setting, but if this were done adjustments had to be made to the front-wheel alignment by altering the length of each drag link. Not exactly a straightforward procedure in a rural situation.

The track setting on the standard-width tractor could be adjusted from 48in (1219mm) to 80in (2032mm) in 4in (101mm) increments. To achieve more extreme settings the front wheel centres need to be reversed. Generally the front wheel tyre size was 6.00 x 16 but 4.00 x 19 could be specified. For the rear wheels the track setting could be varied from 48in (1219mm) to 76in (1930mm), again by increments of 4in (101mm). This adjustment was achieved by altering the bolted relationship between wheel centres and rims.

As another option for the rear wheels, Power Adjustable, Variable Track PAVT-type wheels could be specified. The patent for this design was originally filed by Allis Chalmers but it was certainly used by MF, originating in the USA, from the late 1950s. The principle by which this system worked was that the wheel centre could screw itself in or out of the rim to a limited extent under the power of the tractor! This was achieved by having four short curved steel bars of square section (known as rails) welded to the inner side of the rim in the form of a helix; four clamps were bolted to the wheel centre, each equipped with a locking cam tightened or released by a special square socket spanner. One of these bars on each wheel had holes drilled through to allow clamping type stops to be positioned at predetermined points, giving intermediate track settings on 11.00 x 28 tyres between 52½in (1333mm) and 64½in (1638mm). If the wheels are reversed and changed to opposite sides of the tractor, track widths from 6in (1520mm) to 76in (1930mm) were possible; all settings were 4in (101mm) apart. Rear tyre sizes available as factory fitted could be 10.00 x 28, 11.00 x 28, 13.00 x 24 or 11.00 x 32. These sizes were all offered with three different types of tread: Field Universal, Field and Road, or Grassland/Sand/Road.

Additional weight could be added to the tractor to improve traction or aid stability. At the front a weight frame incorporating a towing clevis could be bolted on, to which up to ten Jerrycan weights could be hung. Each weight was about 60lb (27kg). If the towing facility was needed only eight could be attached. Extra weight at the front could be added with the use of wheel weights. In the case of 6.00 x 16 wheels the weights were of cast iron in two halves and were bolted directly to the inside of the wheel centres, giving an extra 94lb (42kg) per wheel. Weights were available to suit 4.00 x 19 wheels; they were cast in one piece and gave an extra 100lb (45.3kg).

To gain weight on the rear wheels two options were available: either water ballasting with calcium chloride added to prevent freezing, or cast iron wheel weights. The latter were supplied as a kit comprising of four cast weights, each of 108lb (49kg), and special bolts and washers so that one or two weights could be added to each rear wheel centre.

I would like to turn now to some of the changes and modifications that were made to MF135s over their 14 years of production, during which 490,613 units were built in Coventry alone. When the tractor was introduced at the 1964 Smithfield Show the rear fenders were of the shell type, extending down to the foot boards, often known as safety fenders, but by November 1965 an MF135 sales brochure shows a tractor with flat-topped fenders mounted with built-in cast aluminium grab handles also housing the tractor's front side lights. Later, in 1967/68, the fenders were changed yet again, this time to an all-pressed-steel design, thus doing away with the separate cast aluminium part, with a cost saving. Early models had cast aluminium bezels to the headlights, which were built into the radiator grille, but these were replaced in 1969 by plastic mouldings. The early models had both front and rear wheel centres painted red with silver rims, but within less than six months both parts were painted silver. For reasons unknown, the PAVT-type wheels retained their red painted centres to the end.

In 1971 a number of further changes were made. The height of the radiator grille was increased from 12.5in (317mm) to 14in (355mm), allowing the fuel tank capacity to be increased from 8.5gals (38.7 litres) to 10.5gals (48 litres). The steering column was extended due to the bonnet being higher. The front axle was changed from the normal Ferguson type mentioned earlier to a straight, square-section, adjustable type of greater strength. The rope-type oil seal fitted to the

The Coldridge Collections 1976 MF20. This is an ex-military tractor and was bought with only 1400 hours on the clock. When it arrived it was khaki all over, hiding the original yellow finish of the industrial models.

Nearside view of the MF20.

MF20 front view.

This style of Massey Ferguson badge was fitted to all MF construction equipment of this era. It represented a digger bucket.

Nearside view of the standard Perkins A3-152 engine fitted to MF135s – only yellow!

Note the sheet-metal PTO guard of the later models and the chain-operated pick-up hitch.

The speedometer drive taken from the transmission. Later models had this extended PTO control lever.

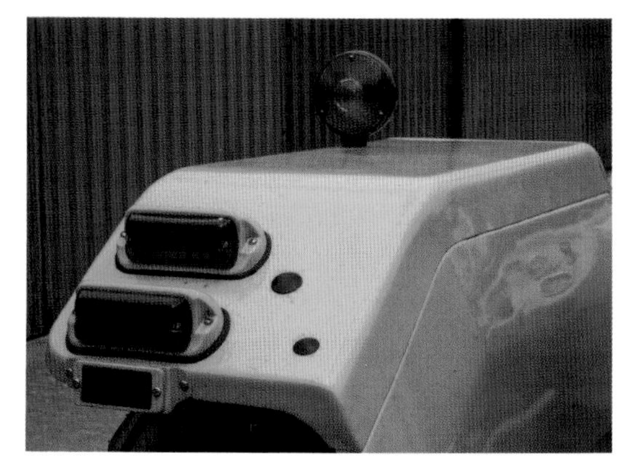

The MF20, being a highway tractor, was fitted with flashing direction indicators, with the tail light in the usual position and the brake light above.

Well-stocked dashboard including a speedometer with odometer, cigarette lighter, indicator switch and repeater. The blanking grommet fills the gap where the temperature gauge for the instant-reverse shuttle gearbox would be, had one been fitted.

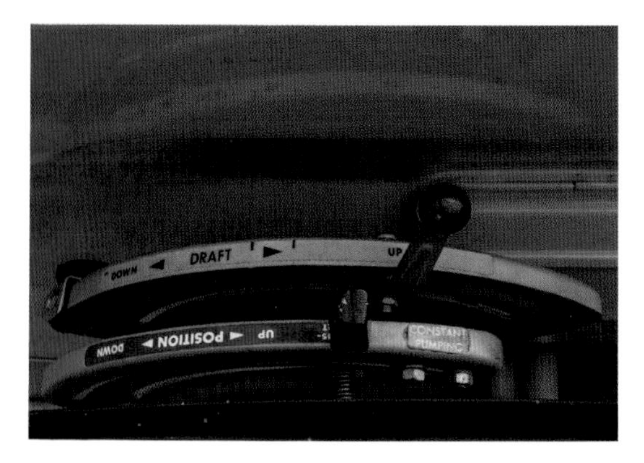

A view of the control quadrants.

The lever in the foreground releases the pick-up hitch.

rear main bearing of the Perkins engine was replaced by a conventional lip-type oil seal of a large diameter to improve sealing. It was in the mid-1970s that the oil bath air cleaner gave way to a dry element type with a service indicator below the dash panel, but the oil bath cleaner was still available for export territories.

The Narrow and Vineyard models were made to meet the needs of farmers specializing in vine, berry, orchard and vegetable production but they did occasionally find application on building or factory sites where their narrowness enabled them to work in confined spaces. Both Narrow and Vineyard models were produced as standard tractors at Banner Lane and were then shipped to Lenfield Engineering (later part of the Ben Turner Group of MF Dealers), which was an MF dealer based in Maidstone (Head Office) and in Ashford and Canterbury, Kent. There they were modified and rebuilt to Narrow or Vineyard specification with the appropriate MF parts shipped over from the MF factory at Beauvais in France, which was the main producer of these special types; presumably the bits left over were sent back to Banner Lane for reuse! Examples of these models in The Coldridge Collection have a commission plate stating 'Made in England'.

In general the Narrow model was very closely related to the standard 135 but had shorter cast trumpet

The dry-element air cleaner (missing its retaining wingnut!)

The air-intake pre-filter extends up under the bonnet adjacent to the battery.

MF135 Narrow Tractor brochure.

housings and half shafts in the rear axle. The track was thus reduced but it was still adjustable within limits in the normal way. Both models were always fitted with shell-type fenders. The lower links had sets in them toward Category 1 ball ends. Another change was to the levelling box handle, which was changed from the normal cranked type to a tommy bar arrangement because of the restricted space by the rear fender. Generally these narrow models were shod with 4.00 x 19 front and 11.2/10 x 28 rear tyres.

The MF135 Vineyard required more drastic modification from the basic model, being made not only narrower but also lower. The reduced height was achieved by the use of smaller front wheels at 5.50 x 15 and rears of 11.2/10 x 24. The dramatic reduction in width was the result of having very short trumpet housings on the rear axle. While the front axle was adjustable within a small range, the steering arrangement required a steering box that operated in the opposite way to normal, ie the swivel arms were turned outwards above the front wheels, thus clearing the bonnet side panels when activated by the drag links.

The rear fenders were of the shell type but made smaller to suit the wheels. The top lift arms of the three-point linkage were a special and shorter forging. The levelling box handle was of the cranked type but

A half-track MF135 performing forestry operations in Sweden.

extended to clear the fenders. Special wheel weights and jacks were produced to fit these vineyard models.

Another area of significant change to the MF135 range came about because of the introduction of the legal requirement to fit a safety cab, designed to strict government specification and approval, which was introduced to the UK in 1970.

The cab initially offered as optional equipment on the early MF135s was of glass fibre on a light steel frame, with good all-round visibility as a result of the generous glazed areas. It even had an electric front windscreen wiper and rear view mirror. "Air Conditioning" could be achieved by removing both doors and/or the roof section – not the environment to be working in should a rollover

This 1978 MF135 belongs to AGCO and is on display at Coldridge.

The quick-detach cab was designed to meet new EU requirement for reduced noise levels in the cab.

Opening side window. Note the direction indicators which were fitted as standard.

Sound-deadening fitted below the steering wheel helped reduce noise from the engine.

Driver's view of the dashboard.

occur. These cabs were made by Duple Coachworks Ltd., the coach builders at Blackpool in Lancashire.

To meet the requirements for rollover protection MF engineers had to redesign and strengthen the gearbox housing to allow the front loader to be fitted with ease. Two types of safety cabs were specifically designed for the 100 Series. The more common version was usually known as the 'Flexi-cab' and was made by Sirocco Engineering, at Tarvin, near Chester. Early Sirocco safety cabs had the windscreen wiper at the bottom and a triangular hand signal flap. Latter examples had the wiper at the top and a rectangular signal flap. The cab was equipped with a safety glass windscreen with an electric wiper. The remainder was clad with a flexible white reinforced plastic material. The right-hand side screen had a clear plastic flap so that the driver could give hand signals. This was the more common cab fitted and, again, fresh air was available: the doors were very easy to remove and the front portion of the cab roof could be hinged back. If the customer required a little more refinement he could specify a cab with the same basic safety structure but fitted with safety glass to the front and sides, and with rigid cladding of sheet steel. This was manufactured by GKN-Sankey, at Telford. All cabs were supplied with ear defenders and a hook on which to hang them when not in use. Both types of cab were compatible with the MF40 front loader, which had a more forward reaching belly plate designed not to foul the cab. Both were built to B.S.4063 1966 and complied with OECD test procedures.

Later in 1976 another piece of EU regulation that affected cab design was the requirement for a lower level of noise, namely a decibel level of less than 90 at the driver's ear. It was this requirement that led to the development of the Quick Detachable top half of the cab, which MF claimed could be removed in less than 10 minutes, special lifting eyes being provided. To achieve a quieter environment, large areas of foam-padded plastic-covered sound-deadening material were fitted over the transmission, the floor, the insides of the rear fenders, the lower part of the doors, the roof panel and most of the dash. These cabs were equipped with flashing direction indicators built into the front and rear of the fenders. An electric windscreen wiper was standard, wired through a plug and socket for use when the top section was removed. The driver had the luxury of an adjustable spring suspension seat fitted as standard after being optional for some years.

More than half a million 135s were built in Coventry alone, and that so many survive today as working tractors or in preservation is a powerful testament to their robust design and quality of build – a tough act to follow.

With a spring-suspension seat and a trimmed cab, this is as luxurious the MF135 got...

...it even had air-conditioning! Opening the lower windscreen gave extra ventilation and allowed access to the battery compartment.

Rear view with doors and rear blind open.

Chapter 4

The MF148 Super-Spec

This Coventry-built tractor was in production from 1972 through 1979, during which time 10,982 units were built. It was an upgraded MF135, but with so many differences and improvements that we need to work closely through the specification and features incorporated in its design to evaluate it fairly.

The engine is again the proven Perkins AD3-152, but built to Series S specification. Although the S specification engine had the same bore and stroke as the Perkins AD3-152 installed in the MF135, it had a totally different cylinder head with a lower compression ratio of 16.5:1, different valve seat angles at 35° instead of 45°, and different CAV fuel injection components.

The rated power output was 47bhp DIN at 2250rpm, with maximum torque at 1400rpm of 131lb/ft. For comparison, the AD3-152 engine fitted to the MF135 had an 18.5:1 compression ratio, was rated at 45.5bhp

This cutaway taken from a Massey Ferguson sales brochure dated January 1972 gives an insight to the MF148's salient points.

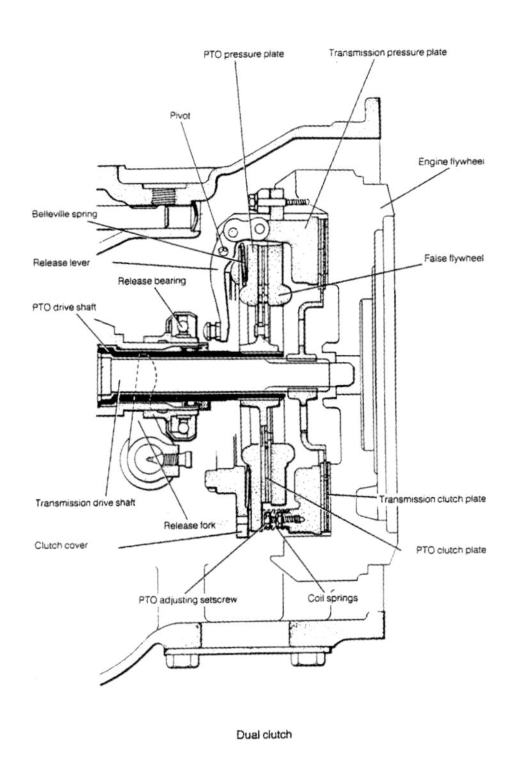

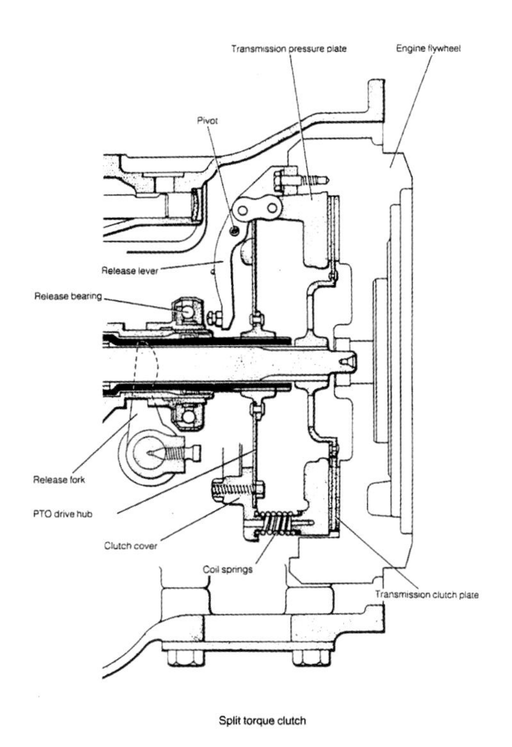

Cross-sectional drawing of an MF dual clutch compared to the split-torque clutch which gave true Independent Power Take Off – IPTO.

at 2250rpm, and had the rather lower torque rating of 119lb/ft at 1300rpm.

The engine auxiliaries are basically the same as on the 135, but power steering was offered as an option, as it was on later MF135s, in which case its pump was bolted directly to the engine, driving off the timing gears. The air intake filter is a dual-element dry type with a warning indicator just below the dash panel. The panel itself is equipped with a comprehensive range of instruments including a tractormeter, fuel level, oil pressure and water temperature gauges as well as an ammeter, hand throttle, lighting switch and engine stop control; a foot throttle to the right side was a standard fitment.

The clutch is a split-torque type of 12in (305mm) diameter, giving a truly independent PTO. It seems appropriate that a description of a split-torque clutch be given here as MF introduced this facility in 1972 on the MF148 as well as on some of the other models in the range. Whereas a normal dual clutch has two 'loose' friction plates, one to disengage the drive to the road wheels and the other to take care of the hydraulics and PTO, a split-torque clutch has only one 'loose' plate. This single 'loose' plate, positioned against the engine's flywheel, disengages drive to the road wheels, but no second plate is used. It is replaced with a steel plate which is bolted to the clutch cover and has a splined boss at its centre; into this is fitted a shaft that takes drive to the hydraulic pump and PTO shaft via its own dedicated hydraulic clutch pack. The main hydraulic lift pump is therefore constant-running. The drive takes place when the clutch pack is pressurised by the operator opening a valve controlled by a small hand lever on the left-hand side that replaces the normal PTO lever. When the drive is disengaged the oil pressure is diverted from the clutch pack plates to a brake band around the body of the clutch pack; this is to ensure that the PTO and the machine it is driving will stop within seven seconds to meet a legal requirement. The hydraulic power to operate this is taken from the auxiliary pump. With this

A Massey Ferguson demonstration Dual Clutch.

George French's well-used 1976 MF148. It's currently wearing David Brown front wheels. The original Massey Ferguson 19-inch rims are very hard to source.

Side view shows the extra length of the Super Spec 148 over the 135.

A close-up of the spacer that increases the wheelbase by 6 inches and allows easier access to the cab.

The cattle seem very interested in this desirable piece of agricultural history.

type of PTO there is no facility to run the PTO relative to ground speed. Tractors fitted with a split torque-clutch are often known as IPTO tractors as they have a truly independent power take-off.

For the gearbox, Multi-Power was standard, providing twelve forward and four reverse gears and thus doubling the number of ratios. As an alternative an eight-speed forward and two-speed reverse could be specified, again incorporating high-low ranges. Behind the gearbox casing was bolted a cast steel spacer of 6in (150mm) in length, a feature of the Super Spec range. This brought about several benefits: i) the longer wheelbase gave much better stability when heavy rear-mounted implements were attached; ii) if front weights were added to the optional weight frame they acted as a counterbalance to the mounted implement when raised, but they also improved traction when the implement was at work because of the extra leverage afforded; iii) the extra length made access to the cab much easier and greatly improved its spaciousness, thus providing a better environment for the driver.

In some cases an MF creeper gearbox was installed retrospectively in place of the spacer; this had its change speed lever on the left-hand side, cranked neatly to fall between the tractor's two gear levers. A creeper gearbox

MF148 dash is unchanged from the 135.

for the MF148 was also available from Four Wheel Traction Ltd, with the change speed lever on the top of the casing. Both creeper gearboxes had a 4:1 ratio, i.e. the same as the planetary reduction gearbox, as standard, controlled by the high/low gear lever.

The rear end of the tractor incorporated several features to ensure that the extra power of the engine, coupled with the ability to handle heavier implements, did not overload it. For example the lift rods on the hydraulic linkage were increased in diameter to ¾in (20mm). The drop brackets to which the inner ends of the lower links are attached were made stronger. Likewise the swinging draw bar was made of 2in (51mm) x 1⅜in (33mm) high carbon steel able to take a maximum vertical static load of 1800lb (816.5kg), the same as the later MF135s. The whole of the rear axle was strengthened to deal with the weight and power: the MF148 had a new design of crownwheel and pinion with larger diameter shafts and bearings as well as changes to the tooth profile. The brakes were of the drum type, of 14in x 2in (356 x 51mm), mounted on the wheel hubs, and could be operated independently or latched together for road work, or used as a parking brake controlled by a hand lever.

The PTO was independent and could be selected by a hand lever on the left-hand side of the transmission housing. Engine speed of 1684rpm gives the standard PTO speed of 540 pm. Alternatively ground speed could be selected (only on eight-speed gearboxes) giving one PTO revolution per 19in (482mm) of forward travel.

The design and function of the hydraulic system followed very closely that of the MF135, but Dual Category interchangeable ball ends were standard, as was a high-capacity auxiliary hydraulic pump on Multi Power tractors only. The steering arrangement was of a similar layout to the post-1971 MF135s, as was the front axle, but made of heavier steel sections. Power steering was an extra. Tyres were 6-ply 600 x 19 at the front and 12.4/11-32 on the rear, on pressed

The MF148 used the Perkins AD3-152 as fitted to the MF135, but built to Series S specification which gave increased power and torque.

Offside view of the S specification Perkins AD3-152. The correct starter motor would have been a Lucas item.

The linkages at the rear end were beefed-up over those of the 135.

steel wheel centres.

The cab was described by MF as a Safety Cab. Apart from the glazed areas and the steel structural members it was clad with white flexible reinforced plastic sections which could be removed, as could the doors. As an extra-cost option rigid steel cladding could be ordered. These cabs did not meet the later Quiet Cab legislation, but a pair of ear defenders for the driver and a hook to hang them on when not on his ears was part of the kit. Whilst on the subject of driver convenience, a cigarette lighter as well as a radio with a set of headphones were optional at extra cost. A nice padded spring suspension seat, a rear view mirror and an electric windscreen wiper were all standard. Later tractors in this range were fitted with Quiet Cabs to meet the 1976 EU regulations requiring a decibel level of less than 90 at the driver's ear level. To achieve this requirement more acoustic material was added to the inside of the cab.

Although the MF148 was nominally available after the 1976 Quiet Cab legislation, the MF550 replaced the MF148 in the UK. Incorporated into the redesign was the Quick Detachable top half of the cab, following the same principle as outlined in Chapter 3. Overall the 148 was a nice useful tractor very much sought after today by MF collectors, one of its much appreciated virtues being high road speed when Multi-Power is used in high. In total MF produced only 10,982 MF148s at Banner Lane. One reason for such poor sales was that in 1976 a MF148, fully loaded with every extra on offer was only £100 less than a MF165 without extras, so you could buy a 4-cylinder 60hp tractor rather than a 3-cylinder 47hp one for just an extra £100!

An MF brochure showing sturdy three point linkage and turnbuckles on the check chains.

Chapter 5

The MF165

The MF165 was the third model in the line-up of the Red Giants introduced to the British public at the December 1964 Smithfield Show in London. It also went into production at MF's Detroit and Beauvais manufacturing facilities at about the same time.

The introductory brochure produced just in time for the Smithfield Show highlights the 165's special attributes thus: "Heavy work or light jobs, this 58 hp model pays both ways. The MF165 gives you single lever control of draft operated implements plus pressure control, the new hydraulic function that brings Ferguson System weight transfer benefits to the use of trailed implements. The MF165 is styled to set the pace for years ahead".

As one would expect, the MF165 could be considered a scaled-up version of the MF135, and we will consider in some detail not only its specification but also the evolutionary changes that were made over its 14 years of production, with 97,744 units produced at Banner Lane alone.

For the UK market the MF165 was generally built as

A photo of an MF165 and three furrow reversible plough from Massey Ferguson's archive collection.

a Standard Clearance model, with a minimum ground clearance of 14 inches (335.6 mm) on 11-32 rear tyres. The rear wheel centres were of a two-piece design similar to those fitted to the earlier MF65s; later models had single-piece wheel centres. A High Clearance model could be ordered which had a minimum ground clearance of 19 inches (482.6mm) on 11-38 rear tyres. Its transmission gearing was modified to give similar road speeds to the Standard Clearance model.

The engine installed at the start of production was a Perkins direct-injection four-cylinder unit designated AD4-203, which had a bore of 3.6in (91.4 mm) and stroke of 5in (127 mm) giving a capacity of 203.5cu.in (3.33 litres) hence the suffix of the engine's designation. With the compression ratio of 18.5:1 it gave a gross output of 58.3bhp at 2000rpm, with maximum torque of 169ft/lb at 1300rpm. The injector pump is a CAV distributor type with a built-in mechanical governor. The injectors were also by CAV and Thermostart is fitted to the induction manifold as an aid to cold starting. An oil bath type air filter was fitted to earlier models but from March 1969 oil bath or dry element types could be specified. This engine was identical to the one fitted to the MF65 MkII, with the exhaust manifold on the left-hand side.

In November 1967 an MF dealer publication makes reference to the MF165 being fitted with a Perkins A4-212 engine, which had the exhaust manifold on the right-hand side – an easy way to recognise the particular engine. There were other differences, this new engine being a downsized version of the A4-236 used in the MF175 and the later MF168. Again it is a direct-injection unit, with a bore of 3.875in (98.4mm) and stroke of 4.5in (114.3mm) – a squarer combination than the AD4-203. The compression ratio is 17.5:1 and it develops 62bhp BS (59bhp DIN) at 2000rpm, with maximum torque of 173lb/ft BS (165lb/ft DIN) at 1250rpm.

The story of the A4-212 is somewhat complicated in that there were two distinct versions of this engine. The initial version, retrospectively called the "low compression" unit, rapidly gained a reputation for poor cold-weather starting, though once started it was a more than adequate performer. After about 18 months, a revised version of the engine was introduced, known as the "high compression" unit, although the actual compression ratio never officially altered. This later version did indeed address the starting problems, but there were some complaints of both lack of power and unpleasant vibration problems. Finally, a further change to the fuel injection overcame both of these issues.

The injection equipment and starting aid were both by CAV, but as the previous paragraph implies each version had its own specific set of injectors and injection pump, as well as differences in the injection pump timing.

Launch brochure for the Massey Ferguson 165.

The electrics were wired negative earth, the standard battery being a 17-plate 96a/h capacity, but for cold climates a 17-plate 125a/h battery was fitted. The starter motor was a Lucas M45G operated by a key switch on the dash panel and wired through a switch controlled by the High/Low gear selector lever, which had to be set in neutral before the starter would turn. For cold climates a more powerful starter would be fitted, a Lucas M50G

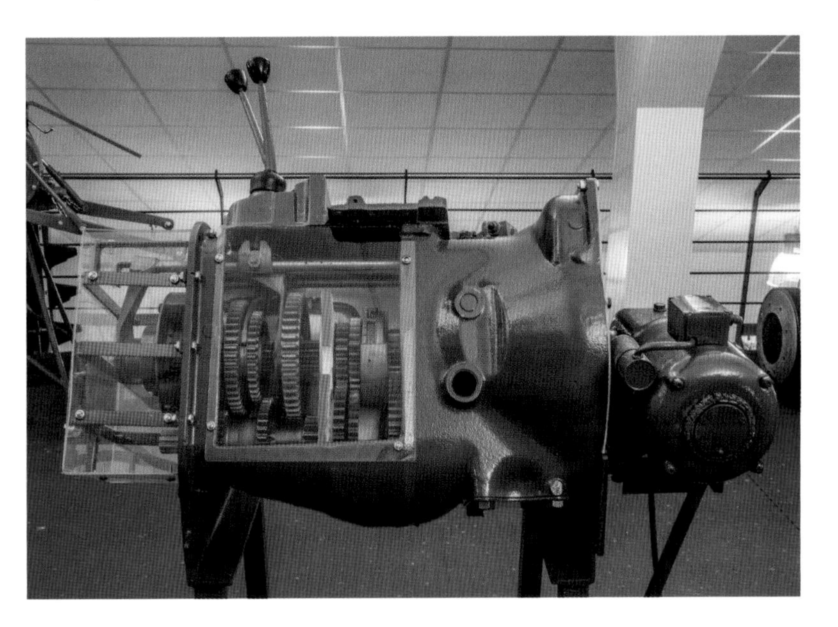

A cutaway and motorised model of the MF165 gearbox equipped with Multi-Power, and the epicyclic unit at the rear end.

An MF165 at work in a Swedish forest. Note the rear mounted Tico hydraulic grab.

(5in diameter as opposed to 4in) – later changed to the M127 when they went metric. A Lucas C40A dynamo together with a Lucas RB108 regulator took care of keeping the battery fully charged.

The clutch fitted to the earlier Standard Clearance models was the dual Auburn self-ventilated type, with a main drive clutch plate of 11in (279mm) diameter, coil spring operated, while the PTO clutch was of 9in (229mm) diameter and Belleville spring operated. The High Clearance models were fitted with what was described as a heavy-duty clutch and a stronger front axle. This type of clutch was eventually fitted across the range of MF165s following the 1971 upgrade; its specification was similar to the early type but the main plate diameter was increased to 12in (305mm) and the PTO plate to 10in (254mm). The dual clutch gave live PTO, which could be selected to drive relative to engine speed or to ground speed. An engine speed of 1685rpm gave the standard PTO speed of 540rpm, while if ground speed were selected 19in (483mm) of forward travel gave one revolution of the PTO.

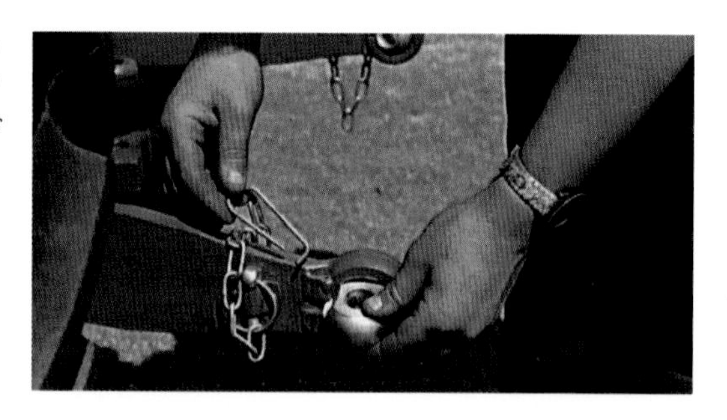

This photograph clearly shows the removable ball ends to give the option of category I or II.

A good photograph of the hydraulic control quadrants on an MF165 taken from a January 1972 sales brochure.

The gearbox was of the sliding spur gear type giving three forward and one reverse gear, compounded by a planetary reduction gear set at the output end of the gearbox which had a ratio of 4:1. As an optional fitment MF's innovative Multi-Power was offered as a factory fitment (see Chapter 3).

The earlier MF165s were fitted with mechanically operated 7in x 4in Girling Ausco dry disc brakes which were housed within trumpet housings adjacent to the transmission case. They could be operated independently for field work by pedals to the right of the driver or latched together for road use; a parking brake was provided as standard, connected to the brake linkage.

In the period around 1970 a number of changes were introduced to the Coventry-built 100 Series tractors. I will quote from notes sent to me by David Walker.

"These changes began on all Coventry tractors with the introduction of the Safety Cab legislation, requiring modification to the gearbox casting to accommodate the front legs of the safety frame. More drastic, and with longer term effect, was the series upgrades on the four-cylinder models, with big changes to the hydraulics, rear axles and brakes, which would continue to be used for a very long time thereafter on later model ranges.

"The revised design work, undertaken at the Maudslay Road Engineering Centre, was in the capable hands of Les Bissett in the case of the rear axle and brakes, and the revised hydraulics were the work of Jim Dean and Bob Yapp, all of whom deserve full credit for their subsequent success.

"In parallel with this, the Banner Lane production facility was also undergoing some major refurbishment, which introduced, in conjunction with these new designs, revised production facilities for the major transmission assemblies. The main feature of these changes was that each gearbox, or axle, now was carried on a guided pathway, with underground cables to steer the assembly trolley, the assembling fitter riding with each one and building it whilst on the move.

"All of this work was completed soon after the summer holiday shutdown in 1971, and the guinea pig, to start things off, was the MF185. The MF165, MF168 and MF188, also incorporating these new features, followed on at the end of the year. That the MF185 was an instant hit is not in question. The combination of well-

A nearside view of Jonathan Lewis' MF165 fitted with the MF safety cab made by Sirocco. This is a later model made in 1974 with silver wheel centres.

tried and already reliable components, in conjunction with the revised axle, brakes and hydraulics, would keep things moving smoothly for the next eight years".

In 1970 MF165s were upgraded in several subtle ways, the most significant being the change to four-plate, 8.7in x 7.4in (222mm x 188mm) oil-cooled disc brakes in the same location as the previous dry brakes. The new brakes gave a much more consistent performance and longer life, and being oil immersed they did not suffer as the dry type could from the ingress of oil. By the way, a tractor fitted with wet brakes can easily be identified by their having an almost square section to the trumpet housing, with four vertical slots cast in it towards the hub end, which indicates the number of friction plates inside.

The steering system used on MF165s was manual but power-assisted steering was available as an option and was frequently specified; in either case 3¼ turns were required from lock to lock. The steering box is a worm-and-roller peg type, which has low frictional losses, with the drop arm pointing downwards on the left-hand side of the tractor. The front end of the drag link is attached to the far end of a short arm that in turn is splined and bolted to a shaft that passes vertically through the front axle support casting. Attached to the lower end of this shaft, again splined, is the lower crank arm, which at

The rear hub of an MF165.

With the door of the MF165 Flexi-Cab open it is easy to see the rather restricted access space. By the way, the cab cover is a genuine MF part recently fitted.

its outer end has two tapered holes to which the track rods are bolted. If power-assisted steering was fitted the double-acting ram and shuttle valve were mounted on the top face of the front axle support casting – a nice tidy set-up, all hidden behind the radiator grille.

The hydraulic pump supplying the system is mounted on the left-hand side of the timing case, incorporating its own built-in oil reservoir, which holds 1½ pints (0.85 litres). This arrangement of the forward part of the steering system meant that it was relatively easy for an MF dealer to fit power steering retrospectively.

American and French MF165s had a different layout in this area, similar to the one used on the MF65 MkII, the Coventry-built MF175 and most of the MF178s. By March 1968 this was changed to the neater and simpler design used on the MF165 as described above.

The hydraulic system fitted to the pre-1971 MF165s follows closely the layout used on the MF135s described in Chapter 3, but being a heavier tractor the rear end was of a stronger build and the hydraulics had a greater lifting capacity at the ball end. The functions available were two-way Draft, Position, Response and Pressure Control, as well as an external services facility. Category 1 and 2 ball ends could be fitted and the top link was the screw-adjustable type. Initially, the lower links had pivoting ends to aid implement attachment but the later versions, with stronger lower links, had interchangeable balls, secured by spring clips. The lower links on the

Note here the right-hand exhaust of the Perkins A4-212 engine.

The MF165 hydraulic system was of a stronger build than the MF135 and had a greater lift capacity at the ball end. Category 1 and 2 ball ends could be fitted and the top link was of the screw-adjustable type.

High Clearance models, as well as their attachment points, were of necessity different.

The lifting capacity at the ball ends of earlier models was 2850lb (1292kg). The four-cylinder piston pump had an output of 3.6gal/min (16.35 litres/min) at an engine speed of 2000rpm or PTO speed of 641rpm, with a relief valve setting of 2500psi. Up to 1.5 gals (6.82 litres) could be taken from the transmission oil sump for external rams.

The upgraded models introduced in 1971 had a lift capacity at the ends of the lower links of 3500lb (1588kg). To accommodate this extra lifting capacity the section of the lower links was enlarged. The linkage pump, again a four-cylinder piston type, delivered 3.1gal/min (14.1 litres/min) at an engine speed of 2000rpm. The relief valve setting was increased to 3000psi (210kg/cm^2). Like other models in the range, if fitted with the optional Multi-Power transmission the lift pump's output could be combined with that of the auxiliary pump to give a much enhanced flow rate to external services such as spool valves or hydraulic motors. The output capacity of the combined pumps at 2000rpm engine speed is 9.4gal/min (42.7 litres/min) at 3000psi (210kg/cm^2), or 12.9bhp.

The fuel tank capacity of the earlier MF165s was quoted as 15 gallons (68.2 litres) but in 1971 it was increased slightly to 17.5 gallons (79.4 litres).

Tyre fitment options were as follows for the pre-1971

The six-stud pressed-steel front wheels were shod with 7.50 x 16 Goodyear Super Rib tyres.

Illustrating the two-piece rear wheel construction using scalloped centres with 10in x 36in rims. The earlier MF165s had three-piece rear wheel centres as seen on the MF65.

The original-type mirror made by Wingard as fitted to both Rigid and Flexi cabs.

The standard-duty front axle with two bolt fixings per side.

Showing the later type oil filter in the Multi-Power hydraulic circuit and also the Gliddon & Squire MF Dealers plate. They are still based in North Devon.

A nearside shot of the Perkins A4-212 engine.

The maker's plate, to prove that it is a Perkins A4-212 engine.

Offside view of the Perkins A4-212 as fitted to the later MF165s, with the hydraulic pump for the power steering bolted to the timing case.

The view from the driver's seat.

standard tractors: front 6.00 x 16 4- or 6 ply with 11.32 4- or 6-ply rear tyres. The post-1971 standard MF165s were normally fitted with 6.00 x 19 6-ply front and 12.4/11-36 6-ply rear. Other options were all at 6-ply rating, 7.50 x 16 front with rears either 13-28 or 14-30.

The early MF165, which had no cab, weighed in at 4400lb (2000kg) whilst post-1971 all-up weights were 5320lb (2403kg) with flexible cladding to the cab or 5445lb (2469kg) with rigid safety cab. Another variation was that while most builds of the early model had an oil bath air filter, the upgraded models were fitted with a dual-element dry air cleaner with a service indicator mounted just below the dash panel; access to the element was by hinging out the metal 165 decal on the top left-hand side of the bonnet.

As regards the cabs, they followed closely those of the MF135 models, so early models had as an optional fitment the fibreglass cab made for MF by Duple, but this was not a safety cab, although the last of the pre-upgrade tractors did incorporate the cab mounting points to meet the safety cab legislation that was being introduced in 1970. The upgraded MF165 introduced in 1971 had by law to be fitted with a safety cab for the UK market. The construction was almost the same as for the MF135. Post-1971 MF165s stood at 92.5in (2350mm) with the flexible cladding while the rigid version was 94.75in (2406mm) to the cab roof.

Among the Coventry-built MF165s were variants that were not sold into agricultural markets. Of this small percentage, some were built up at Banner Lane as highway tractors and painted yellow, while others were sent out as skid units to MF's Industrial Machine manufacturing facility at Barton Dock Road, Stretford, Manchester, to be assembled as "hard nose" diggers and loading tractors. Later in the period covered by this book the Manchester operation was taken over by a management buy-out to become ICM, who went on to adopt a new model numbering sequence including the MF20 (a highway version of the MF135), the MF30C (a highway version of the later Agricultural MF565), and the MF50 loader tractor powered by Perkins A4-236 engine, which sold in great numbers. Since we are dealing in this chapter with the MF165, its highway equivalent was designated MF3165R.

As these models had to meet the Road Traffic Act requirement to have two totally independent braking systems, they were modified by having 14in (355.6mm) drum brakes fitted to the wheel hubs, operated by a hand brake lever on the left-hand side of the driver. The inboard disc brakes were retained as per the agricultural models.

The Instant Reverse Shuttle Transmission is an interesting piece of engineering. The manual shuttle transmission described in Chapter 3 did have definite benefits for the operator when used on loader work but the system was prone to abuse and overheating of the friction clutch. To overcome these shortcomings use was made of the American Funk concept to develop MFs Instant Reverse Transmission. The early MF design was manufactured in Detroit and was first installed in the industrial MF204 tractor in 1959. It continued to be used in MF industrial machines until the introduction of MF Power Shuttle Transmission in 1984. This was a four-speed fully synchromesh unit but by 1987 a fifth speed had been added, directly driven from the engine via a hydraulic clutch pack, thus avoiding the slippage inherent in a torque converter. Both these systems were possibly firsts for MF. For the UK market the Instant Reverse Transmission was built under licence by Turner Engineering of Wolverhampton, with the torque converter made by Borg and Beck. The thinking behind this concept was to reduce operator fatigue, overheating of the normal friction clutch and the general abuse of the mechanical components, for instance from crashing of gears between forward and reverse. Of course with less operator fatigue the expectation was a higher output per man/machine hour. This forward/reverse facility was controlled by the driver's right foot actuating a triple-faced pedal. This functioned thus: the first and slight movement of the right-hand pedal sets an internal valve to direct oil pressure developed by the pump to close the forward clutch pack, while further depressing of this pedal increases engine speed, which in turn causes the torque converter to take up drive and move the tractor forwards. A similar series of events takes place when the left-hand pedal is operated, but as this controls the reverse clutch pack the tractor will travel backwards. The centre pedal simply controls engine speed with no transmission drive but is handy for speeding up the hydraulic functions. This design was the subject of an MF patent and I assure you, the reader, it does work very well especially on loader duties.

This innovative design was tailored to take the place of the normal clutch bellhousing and gearbox as a self-contained unit, and at the same time maintain the structural integrity of the tractor to which it was fitted. In reading this section may I suggest that reference is made to the cutaway drawing above

Replacing the conventional friction clutch was a torque converter with a multiplication factor depending on application of between 2.1:1 and 2.5:1. Mounted immediately behind the converter is a built-in hydraulic pump to provide oil pressure to operate both forward and reverse clutch packs. These are similar in principle to the clutch pack used in Multi-Power transmissions. Next are a set of three gears, one on the input shaft at the top, a reverse idler gear and the third on the layshaft. On the input shaft are the two clutch packs, the first controlling forward movement and the second controlling reverse. Next, and behind the reverse clutch pack, is a gear on the input shaft which meshes with another gear on the layshaft, all of this within the sealed first part of the casing, which is filled with 9.5 litres of automatic transmission fluid (ATF). The other side of this division is the output shaft drive gear, meshing with an independent layshaft which has on it two gears of different diameters that correspond to a selectable pair of sliding gears on the output shaft, the selection of which is controlled by the larger lever mounted on top of the gearbox. Mounted outside the cast steel casing at the rear end is the usual MF epicyclic reduction gear unit, controlled by the shorter gear lever, which had to be set in the neutral position before the starter motor could be operated.

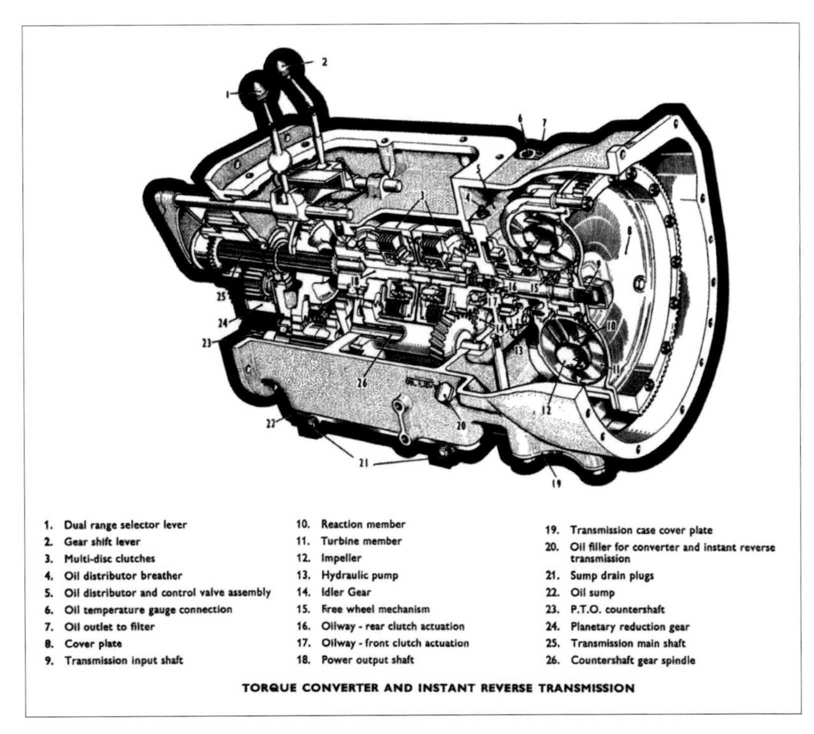

Cutaway drawing of the Instant Reverse Transmission fitted to some industrial MF165s.

The oil in the main two-speed gearbox and the reduction unit was common to the rear axle and hydraulics. The torque converter with its separate ATF had its own oil cooler mounted in front of the water radiator and an additional temperature gauge was fitted on the dash panel to indicate the temperature within the torque converter. The dial of this gauge was marked off in coloured segments to indicate normal operating temperature and a danger area of overheating, which could be caused by an operator trying to use too high a gear. This type of transmission operated reliably in the arduous conditions in which these tractors generally worked.

To round off this chapter it would be fair to say that the MF165 was a most successful machine that gradually evolved over its long production run of 171,441 units manufactured at Banner Lane and is a testament to the people who designed and built this robust and reliable tractor.

Chapter 6

The MF168 Super-Spec

In Chapter 4 we examined the MF148 Super-Spec Model introduced in 1971 along with two others in the same series: the MF168, which is the subject of this chapter and the MF188, which will be considered later.

Like the other models in the Super-Spec line-up, the MF168 benefited from the inclusion of a 6in (150mm) spacer between the gearbox and the rear transmission case. This gave extra length to the wheelbase and not only improved access to the cab, but also gave better stability when handling heavier rear-mounted implements; it also gave a useful gain in rear wheel traction.

The engine chosen to power this model was the

Jon Lewis' MF168 Super-Spec with flexible cladding to the cab.

This view shows to advantage the good access to the cab, when compared with the MF165, that is the result of the 6-inch spacer (see below) in the transmission.

Perkins A4-236 four-cylinder direct-injection diesel, the same engine as installed in the MF175. Its capacity was 236cu.in (3867cc) and it produced 69bhp BS (66bhp DIN) at 2000rpm, with maximum torque quoted as 200lb/ft BS (191lb/ft DIN) at 1250rpm.

There was a 12in (304mm) single-plate clutch of the split-torque type, giving a truly independent PTO driven from the engine via gearing through its own dedicated hydraulically activated clutch pack, which incorporated a built-in brake arranged to quickly bring to a stop the rotation of the PTO shaft and the machine being driven. The standard gearbox was the usual MF six-speed forward and two reverse type but Multi-Power was included as part of the specification, operated by its own high-capacity auxiliary pump, which by the way also operated the PTO clutch pack. The output of this auxiliary pump could be combined with that of the linkage pump to provide for external services. As an alternative to the standard gearbox with Multi-Power, the MF eight forward and two reverse gearbox could be specified, which necessitated the fitment of a dual clutch, but the PTO could be driven relative to engine or ground speed. No auxiliary pump was included in

The 6-inch (152 mm) spacer between the gearbox and the rear transmission case.

Note the optional cast outer front wheel weights and the PAVT rear 14 x 30 wheels.

The Heavy Duty front axle. Note the three-bolt fixing per side. The 'nuts' to the front have a male left-hand thread cut into the outer surface that screws into a tapped thread on the centre solid steel section of the front axle. This ensures that the outer part is firm with the centre section. The bolts from the rear face screw into the 'nut' thus achieving a very solid fix.

The cast steel centre of the PAVT wheel with pressed steel rim shod with 16.9 x 30 Goodyear Super Traction Radial.

The front wheel cast wheel weights attached to the pressed steel centres and shod with 7.50 x 16 Goodyear Super Rib tyres.

The Diverter Valve to the top cover enables oil to be used to feed another output such as a spool valve block or a loader; it also incorporates a return port. When in use it combines outputs of both the lift and auxiliary pumps.

The inner wheel weights are painted Stoneleigh Grey, the outer weights being Massey Ferguson Red as seen in the photo top right of this page.

The rear view showing the Heavy Duty three-point linkage, as fitted to the MF185. It has interchangeable Category 1 and 2 ball ends and is able to lift 1814kg.

At the wheel of the MF168.

this option.

Differential lock was a standard feature as were five-plate oil-cooled disc brakes produced by Girling, 8.75 x 7.4in (222 x 188mm). A parking brake lever was included and the standard steering system was manual, but my guess is that most purchasers would have chosen power steering, which in 1972 would have added an extra £100 to the £2032 (no VAT in those days!) cost of the Standard model with Multi-Power and flexible cladding to the cab.

The hydraulic system incorporated the usual two-way Draft Control, Position and Pressure Control, as well as having three external services tapping points. On Standard specification tractors equipped with Multi-Power the outputs of both the lift pump and auxiliary pump could be combined to give a quoted output of 9.4gal/min (42.7 litres) at 2000psi (140.6kg/cm^2) or 12.95bhp at the same pressure.

The standard front wheels were of pressed steel, shod with either 7.50 x 16 or 6.00 x 19 6-ply tyres. The rear wheels were also pressed steel, the usual tyre size being 13.6/12–36 6-ply, but alternative tyres

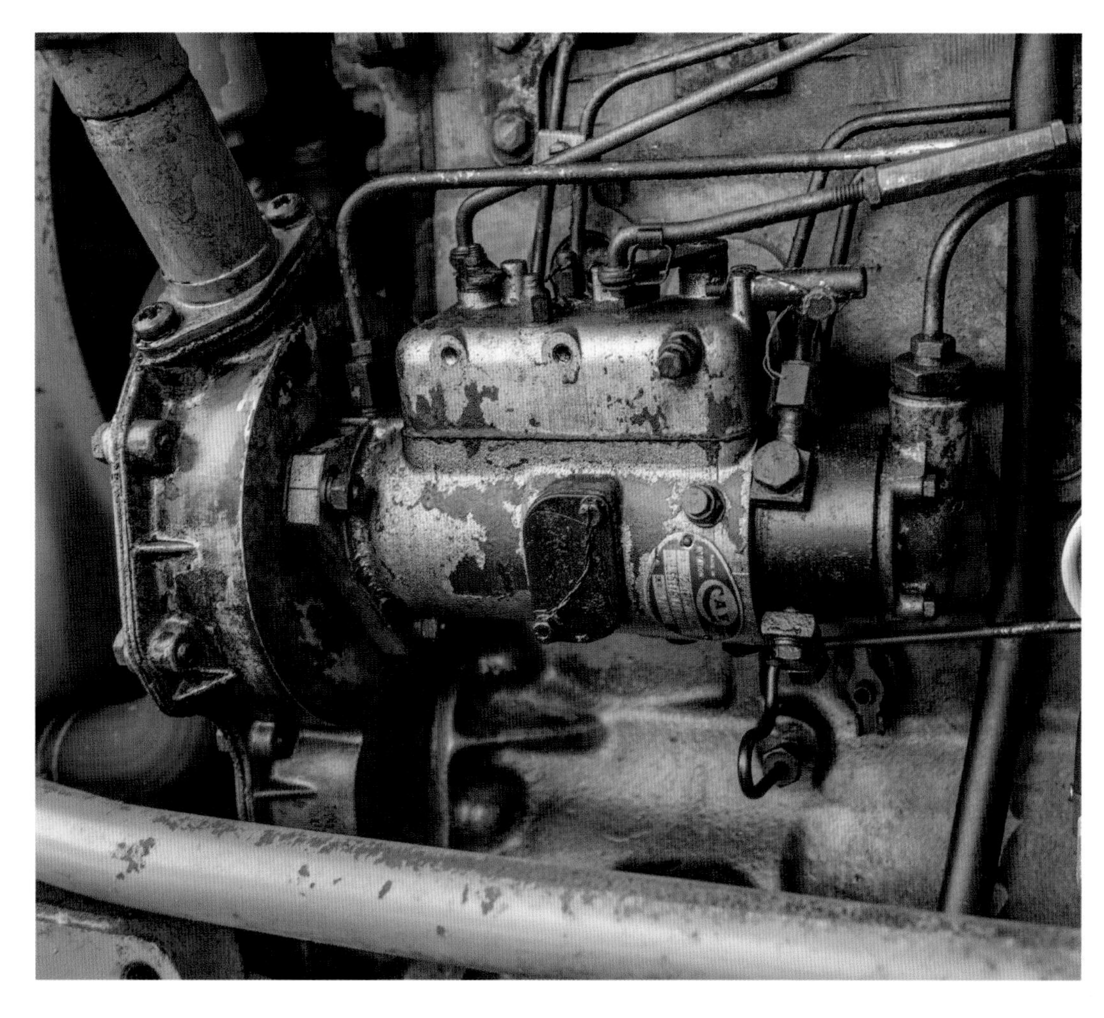

The CAV distributor-type fuel injection pump fitted to this 69hp tractor.

The commission plate of the Perkins A4-236 engine.

were available, 13.6/12-38 or 16.9/14-30, again 6-ply rating.

The air cleaner was the dry dual-element type with a service indicator mounted below the dash.

The standard safety cab for these models was clad with a white flexible material but in 1972, for an extra £50, a safety cab with rigid metal cladding could be ordered. The operator was well provided for with a spring suspension seat, a full lighting set, rear view mirror and ear defenders.

Production of the MF168 at Banner Lane ran from 1971 through to 1979, with 11,173 manufactured.

I would like to round off this chapter on the MF168 by quoting David Walker, who has helped me enormously with this book. "Put simply, the MF168 was one of those machines that did its work extremely well, with a minimum of fuss and bother, almost to the point it went unnoticed except by the discerning few. It was my favourite to drive, being smooth and comfortable, almost one might say, amicable and easygoing." Obviously David had a soft spot for the MF168.

Chapter 7

The MF175, MF175S and MF178

Announced to the British public at the Smithfield Show in December 1964, the MF175 was the largest tractor in The Red Giant line-up. As this chapter will deal with two other models that were really evolutionary developments of the original MF175, namely the MF175S and the MF178, I shall deal with the design and build in general terms, highlighting the differences that were introduced over the production period 1964-1972.

The first sales brochure produced by MF for the 175 launch boldly states, "This 66hp tractor is the new class leader in performance, value for money and features". I feel from personal experience that I can endorse that claim. A nearby farmer was an early purchaser of an MF175 and word quickly got around that this gentleman had bought a monster tractor. As he was a customer of mine I immediately phoned him and asked if I could come and have a look at it, which I did. Yes, it did appear enormous but that was in February 1965 and we were not then used to 150hp tractors as we are today.

The engine used to power the 175 was the Perkins

The MF175 was the largest of the Red Giant line-up.

The Coldridge Collection's MF175. This tractor was originally purchased by the Berkshire Agricultural College.

direct-injection A4-236, with a bore of 3.875in (98.4mm) and stroke of 5in (127 mm), giving a capacity of 236cu.in (3867cc) hence the engine's suffix. The compression ratio was 16:1 and maximum power was 66.4bhp at 2000rpm, with maximum torque of 190lb/ft (26.2kg/m^2) at 1200rpm. The CAV injector pump was a distributor type with built-in mechanical governor; the injectors and Thermostart cold starting aid were also by CAV. The fuel lift pump was by AC Delco, with a hand priming lever followed by two fuel filters with replaceable elements, the primary filter casing incorporating a glass sediment bowl also known as an agglometer. The air filter was of the oil bath type and a standard fitment across the three models described in this chapter.

The MF175S was powered by the same Perkins engine as the MF175 but the MF178 was fitted with the Perkins A4-248 which was an upgraded version of the A4-236. With an extra 12cu.in (197cc) it gave 6.1bhp more at 2000rpm with a rise in torque to 213ft/lb (29.3kg/m) at 1300rpm, the quoted power output being 72.5bhp at 2000rpm. This model was introduced at the Smithfield Show in December 1967. Expressed in percentage terms the hp was increased by 6% but more importantly the torque went up by 14%.

The clutch fitted to the MF175 was an Auburn ventilated dual type with a main clutch plate diameter of 11in (279mm) coil spring operated, while the PTO clutch was of 9in (229mm) diameter and Belleville spring operated. The clutches of all three of the models dealt with in this chapter are of the same dimensions and specification. Whilst the MF175 and MF175S continued with the 11x9 clutch, from August 1968 the MF178 was fitted with the larger 12x10 dual clutch.

At the introduction of the MF175 in December 1964

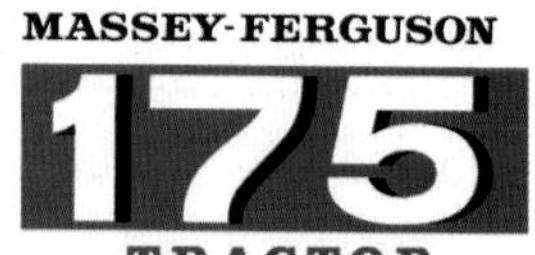

The front cover of the MF175 launch brochure dated August 1964, printed well ahead of time!

Nearside view of the MF175.

two gearbox options were available: a basic model with the usual six forward and two reverse speeds, the gears being of the sliding spur type and accompanied by the usual planetary 4:1 reduction arrangement controlled by a secondary gear lever. The Multi-Power System was offered as an optional factory fitment at additional cost.

When the MF178 was launched it was only available with a six-speed forward and two reverse gearbox, with the option of fitting Multi-Power, but with the introduction of the Coventry-built Super-Spec tractor range in 1971 an eight-speed forward and two reverse 'box became an option.

Offside view of the MF175's Perkins direct-injection A4-236 engine.

The MF175S was produced for a very limited number of markets, notably Scandinavia (is that what the S signified?), going, as far as is known, to Sweden, Norway and Finland. The upper part of the cab was painted silver and a cab heater with defrosting ducts was standard.

The steering arrangement on the MF175 followed that used on the earlier MF65, with the characteristic high-line drag link and pedestal. This axle is easily identifiable by the front support casting having an inverted 'U' shaped mounting pad on each side for loader fitment, whereas the MF165 had square pads. This type carried on being used until about half way through production of these models, when the MF165 version was adopted with the drop arm facing downwards and the drag link alongside the engine at starter motor level.

For some reason the MF175S had manual steering as standard with the option of power assistance as an extra, but at 2355kg it was such a heavy tractor that I expect most buyers paid the extra cost, which at the time was about £100.

The front wheels of MF175 and MF178 had cast steel centres while the MF175S had front wheels of

What appears to be an MF175 operating in Sweden towing a Saab Viggen.

pressed steel throughout. The rear wheels of both the MF175 and MF178 were of the PAVT type with cast centres. Various pressed steel options and tyre sizes could be specified across the range.

Turning now to the hydraulic systems fitted to these three different models, they followed the same layout as on the MF165 described in the previous chapter. On the MF175 two options were made available, with or without Pressure Control. All were fitted with heavy duty Category 2. The linkage pump was the four-cylinder scotch yoke type which gave a maximum working load on the lower links of 3375lb (1530 kg). Its output was rated at 3.6gal/min (16.37 litres/min) at an engine speed of 2000rpm. Three tapping points were provided for external services. If the tractor was fitted with the optional Multi-Power transmission then the auxiliary gear-type pump, rated at 5.8gal/min (26.5 litres/min) at 2000rpm engine speed, could be combined with that of the lift pump to produce a flow rate of 9.4gal/min (42.7 litres/min) at 2000rpm, which expressed in terms of horsepower would be 13.2bhp, with 8.2bhp being derived from the auxiliary pump. It may be worth stating here that the relief valve setting on the linkage pump is 2500psi (176kg/cm^2): while the relief valve setting on the auxiliary pump is 2000psi (140kg/cm^2). The MF175S and the MF178 offered a similar system but on the MF178 pressure control was standard and the maximum system pressure was 3000psi (210.92kg/cm^2).

All these models were fitted with mechanically operated 7in x 4in (177mm x 101mm) double dry disc brakes incorporated within the rear axle trumpet housing adjacent to the transmission case. These brakes therefore applied their braking force to the half shafts, which ran faster than the wheel hubs. A parking brake was standard across the range, connected to the foot brake linkage, except for certain markets such as West Germany where a dual braking system was a legal requirement even on agricultural tractors. To meet this need MF followed the same approach as was applied to Highway tractors used in the UK, namely to install at each rear road wheel hub a 14in (255mm) diameter drum brake activated by the parking brake lever.

This photograph was taken from an MF Product Information booklet dated 1967 and shows to good effect the use of Pressure Control when pulling heavily laden four-wheel trailers.

Again the cab design and manufacture followed the same basic outline as detailed in Chapter 3, with only the very late MF178 models actually being fitted with safety cabs, supplied either by Sirocco or by GKN Sankey.

MF175s built at Coventry but destined for export to Germany were in fact badged MF177. This was because the number 175 related to a German law known as Paragraph 175, dating from 1871, which strictly prohibited homosexuality. Thus 175 could be used in a socially derogatory way; this law was repealed in 1994.

An impressive total of 53,107 MF175, 177 and 178 units were built at Banner Lane between 1965 and 1971, while only 7362 MF175S tractors were built 1968-1971.

In Germany, the MF175 was actually badged the MF177.

Chapter 8

The MF185

When introducing the 100 Series range in 1971 MF claimed their 185 tractor was unbeatable for value and work power, and went on to state, "There are plenty of exciting options but this basic machine can supply more useful power and performance for its price than any other 75 hp tractor in Britain".

Let us consider the specification of this basic model in the higher horsepower range. A 1972 MF list gives a price of £1956 as well as the prices of optional fitments, but we'll take the basic model first.

The engine chosen to power this tractor was the Perkins A4-248 four-cylinder direct-injection diesel with a capacity of 248cu.in (4.06 litres), the bore being 3.98in (101mm), stroke 5in (127mm). Quoted output was 75bhp BS (71bhp DIN) at 2000rpm with maximum torque of 219lb/ft (28.2kgf/m^2) BS (208lb/ft DIN) at 1300rpm. The injection pump was the distributor type by CAV, who also made the Thermostart to aid cold starting. The air filter was a dry dual-element unit with service indicator. The clutch was a dual type with a 12in (305mm) main clutch and a 10in (254mm) PTO clutch, giving live PTO related to engine speed or ground speed. The gearbox was an eight-speed unit and

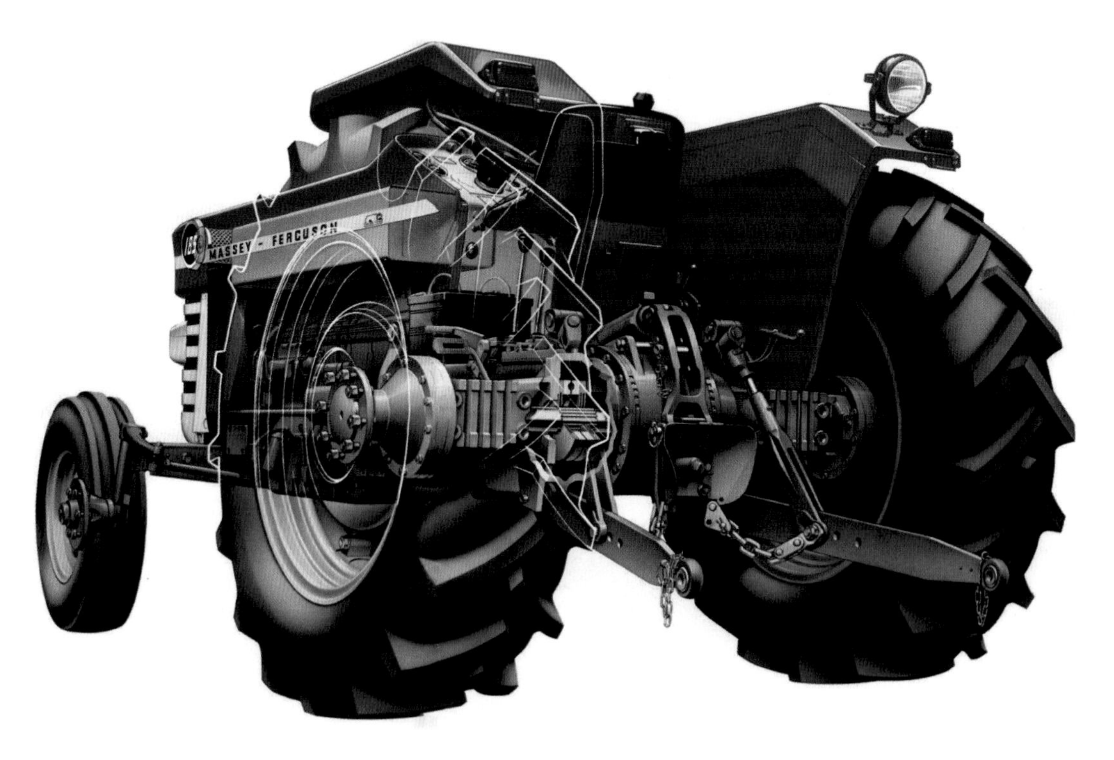

This cutaway is taken from the launch brochure of the MF185, dated January 1972. It says it all!

An archive photograph taken from the same brochure showing the MF37 Cultivator hitched to a MF185.

The linkage was heavy duty Category 2 three-point linkage with interchangeable ball ends, and the top link was screw-adjustable with a lift capacity of 4000lb.

the rear axle incorporated a differential lock.

The hydraulics incorporated Draft, Position, Pressure and external services control. The linkage was heavy duty Category 2 three-point linkage with interchangeable ball ends, and the top link was screw adjustable with a lift capacity of 4000lb (2064kg). The basic hydraulic system had a four-cylinder scotch yoke lift pump delivering up to 3.1gal/min (14.1 litres/min) at 2000 engine rpm. One and a half gallons (6.81 litres) was available from the transmission case to supply external rams.

The brakes were five-disc oil-immersed of 8.75 x 7.4 in (222 x 188 mm), operated by pedals or a parking brake lever. The steering was manual. The front wheels were pressed steel with detachable inner weights and were shod with 7.50 x 16 6-ply tyres. The rear wheels were also of pressed steel, and fitted with 13.6/12 x 38 6-ply tyres. A front weight frame and weights were an optional extra.

The MF185 in standard form was supplied with a flexibly clad safety frame, full instrumentation, ear defenders, rear view mirror and a spring seat. For those users needing additional refinements MF offered various packages: for an extra £50 one could have a rigid cab, while if power-assisted steering was thought to be a good investment that would cost an extra £100. Available as an alternative to the standard eight-speed transmission the MF twelve-speed set-up incorporating Multi-Power would have added £115 to the basic price. You could also have a more versatile hydraulic system with a high-capacity auxiliary pump which could deliver 6.3gal/min (28.63 litres/min) with 8.8bhp available; there was also the facility to combine the outputs of both lift and auxiliary pumps to give a total of 12.95bhp. Alternative rear wheel and tyre sizes were offered, 16.9/14 x 30 or 16.9/14 x 34

During the period of production, starting late in 1971 and ending in early 1979, a total of 40,096 MF185s were built at Banner Lane. As a comparison the MF188 Super-Spec model, the subject of the next chapter, had a production total of 21,333 during the same time span at Coventry, so clearly it was a wise decision by MF to market the keenly priced but somewhat basic 75hp tractor that the 185 was.

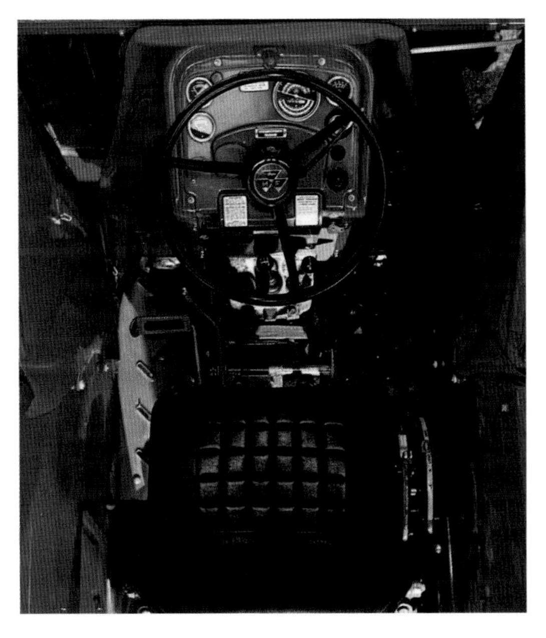

At the controls of the MF185.

Chapter 9

The MF188 Super-Spec

The most sophisticated model in the 100 Series was the Super-Spec MF188, which was produced at both Coventry and Beauvais. Most of the UK production was sold to the home market or exported, mainly to the Scandinavian countries.

The engine powering this model was the Perkins A4-248, the same as used in the MF185. This engine had a bore of 3.98in (101mm) and a stroke of 5in (127mm), giving a capacity of 248cu.in (4.06 litres). With a compression ratio of 16:1, quoted output at 2000rpm was 75bhp BS (71bhp DIN), with maximum torque at 1300rpm of 28.2kgf/m^2.

This cutaway drawing of the MF188 is taken from a January 1972 sales brochure and gives a good overview of the tractor.

Being the top of the range it was comprehensively equipped – those with cabs even included ear defenders in the specification. Like all the models in the Super-Spec range it benefited from the inclusion of a 6in (150mm) spacer between the gearbox and the rear transmission housing, thus not only giving a more spacious safety cab but enhancing the stability of the tractor, especially when handling heavy rear-mounted implements.

The 12in (305mm) clutch was of the single dry split-torque type coupled to the normal MF six-speed forward and two reverse gearbox together with Multi-Power, which was a standard fitment. The PTO on tractors with this build was truly independent, being controlled by its own hydraulic clutch pack. As an alternative the MF eight-speed gearbox could be ordered, in which case a dual clutch was fitted, thus providing a PTO that could be driven relative to engine speed or ground speed. A tractor built to this specification was priced at £85 less than the more usual model, which with a flexibly clad safety cab cost £2173. It should be noted, as with almost all MF100 Series models, that the dual clutch was fitted to tractors destined for export to Scandinavia and other cold-climate countries to aid starting.

The hydraulic lift system had two-way Draft, Position, Pressure and Response control as well as hydraulic tapping points for external services. The lift linkage had only Category 2 attachment points, but the lift capacity was 16% greater than that of the MF185 at 4600lb (2064kg). Although the linkage looked the same, the geometry had been revised to give the extra lift. The top link was of the telescopic screw

type. Driving the hydraulic system was a four-cylinder scotch yoke piston pump, constant running, with a maximum delivery at 2000rpm engine speed of 3.1gal/min (14.1 litre/min) with a pressure of 3000psi (2109 kg/cm^2). The gear-type auxiliary pump operated the clutch packs of both the Multi-Power and the PTO as well as adding to the output for external services. This pump had a rated output at 2000rpm engine speed of 6.3gal/min (28.6 litres/min), or 8.8hp of hydraulic power. Maximum pressure was rated at 2000psi (140.6kgf/cm^2). The outputs of the two pumps could be combined to give 9.4gal/min (42.7 litre/min) at 2000psi, or 12.95 hydraulic horsepower.

The brakes were five-plate oil-cooled discs of 222 x 188mm diameter, operated mechanically by two pedals, independently for field work or latched together for road use; a parking brake lever was provided as were a differential lock and power-assisted steering. The air cleaner was of the dual-element dry type with a service indicator mounted below the dash panel.

According to an MF publication dated December 1971 the safety cabs were generally clad with a flexible reinforced white plastic material, but at an extra cost of £50 metal cladding could be specified. Towards

Jonathan Lewis' MF188 was fitted from new with a Duncan safety cab, with big doors and an upright rear stance. This cab was priced halfway between the cost of an MF Flexi-Cab and the more expensive MF Rigid Cab (good marketing on the part of Duncan!)

The open door shows to good effect the tractor's extra length and the wide opening.

This tractor is a special order built with an eight-speed gearbox. Most 188s were built with six speeds and Multi-Power

The arrangement of the front-opening windscreen of the Duncan safety cab.

A close up of the offside front pressed steel wheel fitted with inner and outer cast weights.

The PAVT rear wheels with cast steel centres and pressed steel rims shod here with 16.9/14 x 34 Goodyear Super Traction Radials.

The trim concealing the windscreen wiper motor and drive also provides a fixing point for the interior light, as well as a location for the wiper switch and the direction indicator switch and repeater.

The resilient offside mounting point for the cab, a quick-release concept.

Left-hand side of the Perkins A4-248 engine.

The spring suspension seat fitted as standard to Super-Spec models.

the end of production, to comply with legislation, all tractor cabs were metal clad and fitted with additional sound-deadening material. These cabs were the Quick Detachable design as outlined in Chapter 4 on the MF148 tractor. Flexible cladding was discontinued. For driver convenience a foot throttle, rear view mirror and a spring suspension seat were all provided as standard equipment, likewise a full lighting kit including a ploughing lamp. The dash panel featured a tractormeter and ammeter along with fuel, oil pressure and water temperature gauges.

The standard wheels were pressed steel at the front, fitted with both inner and outer cast wheel weights and shod with 7.50 x 16 6-ply tyres. The rears were Power Adjustable Variable Track (PAVT) with cast centres. The standard rear tyre size was 12-38 6-ply, optional sizes being offered of 14-34 or 15-30, all in 6 ply-rating.

The Beauvais factory produced a four-wheel-drive version of the MF188 starting in 1972. This is possibly

An offside view of the Perkins 75hp A4-248 engine. Note the primary fuel filter with a transparent sediment bowl and the hydraulic pump for the power steering.

The ball ends were not interchangeable and were only used on the MF188.

The heavy duty front axle with the three-bolt fixing per side as described in Chapter 6.

The heavy duty lower links of 4½in x ¾in alloy steel; the lift rods are 1 inch in diameter with an optional "float" facility.

Note the lift rods for the pick-up hitch.

The heavy duty rear end with a lifting capacity of 2064kg at the Category 2 ball ends.

This shows the optional heavy duty front weight frame, able to accommodate up to eight special 90lb (40.8kg) weights as used on the MF1080.

The MF188 also incorporated a 2½-inch spacer between the front tombstone and the engine to give enhanced stability.

MF's first foray into four-wheel-drive tractors. Prior to this MF tractors were sometimes converted to 4WD with equipment designed and provided by outside manufacturers. The front tyres fitted to the MF 4WD tractors were 11.5/10 x 24 with the drive to the front axle engaged by a hydraulic multi-plate clutch, thus giving some protection against the overloading of the front axle.

The Coventry-built MF188 with a metal cab weighed 7082lb (3220kg). Just by way of comparison, a standard French-made MF188 with a cab weighed in at 2858kg, while the French-produced 4WD without the cab weighed 3400kg, an extra 542kg!

The output of MF188s from Banner Lane through the period of production 1971 to 1979 amounted to 21,333 units: a just confirmation of their popularity with farmers wherever in the world they were sold.

From the launch brochure – the MF188 hitched to an MF41 plough.

Chapter 10

Implements of the 100 Series

If we consider the implements listed in the MF Salesman's Pocket Catalogue of 1972 we find a range of implements designed and engineered to team up with the higher-horsepower tractors in the 100 Series range. Alongside these some of the smaller implements from the earlier era continued to be marketed for farmers using the smaller tractors produced at this time.

At the risk of being boring it seems appropriate to set out the whole range of machines, excluding balers and combine harvesters, that are listed in that 1972 publication, with an asterisk against the implements dealt with in my previous book *Massey Ferguson 35 & 65 Models In Detail,* which will not be covered here.

David Walker reminded me that from the late 1960s MF relied much more on buying in implements from a wide assortment of suppliers, both UK-based and elsewhere. This included the French Huard company for the plough range, Langeskov and Mads-Ambi for cultivators, Bomford for the chisel plough and Weeks for trailers, plus others. To balance this, there was still design responsibility for certain other implements, such as the MF35 and MF40 loaders, the MF11 spinner broadcaster and, above all, the superb and highly acclaimed MF30 and MF130 seed drill ranges. There was, and remains, a feeling among some ex-MF staff, operators and former dealers that something was lost when there was not the same care and attention paid to plough design, carefully tailored to the tractor, as had been the case in earlier times.

Starting this list with implements for Cultivation that were available from Massey Ferguson within this period we have:

- MF41, mouldboard ploughs with rigid beams
- MF41, ploughs with spring release beams
- MF46, Two Furrow Reversible Plough with rigid beams
- MF47, Two Furrow Reversible Plough with spring release beams
- MF30, Three Furrow Reversible Plough with rigid beams
- MF32, a heavy duty version of the MF30 plough
- MF34, a heavy Three Furrow Plough with spring release beams
- MF26, described as a Four Furrow Reversible Plough for large acreages and big tractors
- MF86, a semi-mounted Four or Five Furrow Plough
- MF24, a Heavy Chisel Plough
- MF23, a Coil Tine Cultivator available in five different widths
- MF39, a Cultivator with spring-loaded tines manufactured in seven widths
- *MF738, a Tiller with spring-loaded tines as a safety feature available in three widths
- *MF770, a heavy duty Spike Tooth Harrow
- MF52, a heavy duty Disc Harrow
- MF28, a Disc Harrow available in four widths

Under the heading Fertilizing and Planting we have:

- MF10 or MF11 (very similar), a Mounted Spinner Broadcaster
- MF12, a trailed Spinner Broadcaster with a large hopper capacity
- *MF19, a Manure Spreader available in two sizes
- MF29, a Seed Drill with built-in Fertilizer Hopper made in three widths
- MF34, a Multi Purpose Drill made in two widths

Lastly, implements for Mowing, Loading and Transport:

- MF20, a 3-ton Low Loading Trailer
- MF21, a 3½-ton Tipping Trailer
- MF22, a 4½ ton Tipper
- MF23, a High Level Tipping Trailer
- MF24, a 6-ton Tipping Trailer
- MF25, a Twin-Axle 7-ton Tipper
- *MF702, a Linkage-Mounted Transport Box
- *MF721, a Multi-Purpose Blade
- MF40, a Front End Loader designed to be suitable for all the Coventry-built 100 Series tractors, with use of appropriate brackets
- MF32, a Rear Mounted Reciprocating Mower
- MF51, an Offset Rotary Mower

From this listing it will be clear that the range of implements was tailored to match the quite wide range of tractors making up the 100 Series.

Before dealing with the ploughs in detail it would seem prudent to list and briefly describe the range of bases that MF made available to accept different types of mouldboards, ploughshares and bar parts for its range of ploughs.

The L base, described as good for autumn ploughing of heavy stubble, leaves an unbroken furrow slice and works well down to 9in (228mm).

The N base is of the digger type, whose share cuts the whole width of the furrow slice. Very good for seed bed preparation and burying trash and growth, these work well down to a depth of 12in (304mm).

The C base is a deep digger type for use with relatively short mouldboards with a pronounced concave curvature. Single and two-piece shares are available for this type, which works well down to 14in (355mm).

The V type base is for the Bar-Point and has much the same characteristics as the N type. The Bar-Point is particularly useful on stony ground, and works well down to 12in (304mm).

The W base has characteristics between general purpose and semi-digger. It achieves a well-packed furrow slice and good coverage of trash; the share is of a two-piece design and a replaceable mouldboard shin is fitted.

The HDS (heavy duty speed) base has a longer share and mouldboard than the N base; in addition the mouldboard has less curvature and slope. This type has an outstanding ability to bury trash without "throwing" the furrow slice when used at the high speed at which this base is capable of operating.

Just to add a bit of confusion some MF plough publications describe the L base as general/ley and the H type as general purpose.

With the wide range of base types available on most of the MF ploughs of this period it is not surprising to find two types of coulters being offered on most plough models, being disc with a single arm or knife.

Four types of skimmer could be specified, known as ZC disc, ZD general purpose, ZF manure and ZU universal.

Listed as optional accessories for some models of MF ploughs available in this period was an adjustable control wheel to assist when carrying out shallow ploughing. Also available as an extra on some of the larger ploughs was the Transport Wheel with a pneumatic tyre, the purpose of which was to take the strain off the hydraulic system when travelling at speed on the highway; it could also be used to perform the duties of a depth wheel. I wonder what Harry Ferguson would have thought of that!

Let us now consider the range of ploughs designed and marketed by MF in this era. First, the MF41, a fixed plough with an extra strong backbone made up of forged steel beams that were rigid. This plough could be supplied as a two-furrow in 10in, 12in and 14in width settings, as a three-furrow in only 12in and 14in settings, or as a five-furrow only in 12in. The linkage pins on this range of ploughs were as follows: two-furrow Category 1, three-furrow Category 1 and 2, four- and five-furrow Category 2 only. For the coulters there was a choice of single arm discs or knife; the skimmers could be ZC disc, ZD general purpose, ZF manure or ZU universal. Various bases could be specified as follows: Y – semi-digger/deep digger, H – general purpose, L – general/ley. A transport wheel

The MF41 Four Furrow Plough with rigid beams.

The MF41 Four Furrow Spring Beam Plough.

was optional on the four- and five-furrow models. A depth control wheel could be supplied as an optional extra on the three-, four- and five-furrow models.

Next we have the MF41 with Spring Release beams. These were built more or less to the same specification as the range just described but as their name indicates each beam was spring loaded, thus avoiding any damage to the frame should a serious obstruction be encountered.

The MF30 was a tough three-furrow reversible plough with mechanical turnover or indexing and Category 2 linkage attachment; the furrow width could be set to 12in or 14in. The headstock not only incorporated the turnover mechanism but also an adjustable spring-tensioned safety breakaway device that allowed the plough to "knuckle" if it struck an obstruction. The pivoting cross shaft arms allow easy coupling to the tractor. A transport wheel was fitted as standard, which of course could be used as a depth control wheel. The bases available were Y, H and L, and as with most other ploughs in the MF range they could be equipped with knife or single arm disc coulters together with the usual range of skimmers.

The MF34 Reversible plough was very similar in specification to the MF30 just outlined but built to a slightly cheaper price. This was achieved by deleting the headstock release system but providing a safety factor with spring release beams.

The MF32 Three Furrow Reversible plough was in reality a heavy duty version of the MF30 plough but

The MF30 Reversible Plough being pulled by an MF185.

The MF34 Three Furrow Reversible Plough.

The MF32 Three Furrow Reversible Plough.

was equipped with a hydraulically operated turnover/indexing mechanism. The main frames were of forged steel for extra strength and a transport wheel was fitted as standard. MF made it clear that these ploughs were intended to team up with tractors of over 70hp, but when extended to a long clearance version they could be successfully matched to tractors of over 90hp.

The MF26 Reversible Four Furrow Plough.

The MF26 plough was described as a "massive four-furrow reversible for large acreages and big tractors". Linkage attachment was Category 2 only and the furrow width was set at 13in (330mm). This model could be supplied with either rigid or spring release beams; if the latter, only knife coulters could be installed, but for the rigid frame version either single arm discs or knife coulters could be fitted. The indexing mechanism in the headstock was hydraulically operated. The usual range of bases could be specified i.e. Y, H and L. A transport wheel was a standard feature and was readily adapted as a depth control wheel by the installation of a special bracket.

The MF46 Reversible Plough was made only as a two-furrow model with width settings of 12in (304mm) or 14in (355mm); the linkage attachment was Category 1 or 2. This model was designed to team up with the smaller tractors in the 100 Series. The usual range of bases and coulters could be fitted. The indexing of this plough was achieved by a mechanical mechanism.

Moving now to the last mouldboard in the MF line-up we have the MF86, a semi-mounted machine imported into the UK from Canada. Being semi-mounted, the pulling tractor, if equipped with the MF Pressure Control System, benefited from the draft forces of the plough aiding wheel adhesion. On being lifted out of work the front end of the plough was raised by the tractor's three-point linkage while the rear end was raised by a remote hydraulic ram lifting the rear steering wheel. The steering effect was achieved by a linkage from the front pivoting hitch attachment member; this enabled the ploughman to have a fairly narrow headland for such a large plough. It was generally built with four or five furrows set at 12in (304mm) but a sixth furrow could be added. These ploughs had a spring-loaded trip mechanism built into each base support, factory set at 6800lb (3090kg) plus-or-minus 10%, but they were user adjustable. The disc coulters were 18in (457mm) single or double arm type with either general purpose or scotch skimmers. The bases that could be specified were L, Y, N or HDS (heavy duty speed).

In my own limited experience of using a five-furrow version of this plough behind an MF175 with Multi-Power it certainly pulled the rear wheels firmly into contact with the ground as well as doing a good job of ploughing.

This image taken from an MF sales brochure dated 1966 of the MF86 Semi-Mounted Five Furrow Plough; note the remote double-acting hydraulic ram that raises the rear trailing wheel.

The MF24 Chisel Plough, to my eye, is really a heavy duty cultivator. This implement was built in a range of widths so that it could be matched to the power of the tractor to which it was coupled. The smallest was 78in (1981mm) with Category 1 and 2 linkage attachment, while the three larger models had just Category 2 linkage. The widths of these three larger machines were 96in (2438mm), 120in (3048mm) and 168in (4267mm). These robust Chisel Ploughs were built with three heavy steel box sections forming the lateral members of the frame. Bolted to

The MF24 Chisel Plough being put through its paces.

The MF39 Cultivator at 216in (5486mm) wide in transport position.

The MF52 heavy duty Disc Harrow.

these were the tines, made of fabricated steel pressings terminating with a steel shoe with a breakaway shear pin to protect the double-ended chisel share, which was a 3in (7mm) x 5/16in (7mm) section of hardened steel. The MF24 was, in fact, a re-badged Bomford "Super Flow" chisel plough.

As an option, subsoiling tines could be fitted, designed to transmit extremely high loadings, the wearing surfaces being of Ni-hard steel with the underside of the parts chilled for longer life and designed to be self sharpening. Also as an extra-cost fitment, two rear depth wheels could be installed when using the implement as a subsoiler or for use behind a tractor not fitted with draft control. For some reason these depth wheels were fitted as standard on the two larger implements.

In the range of disc harrows the MF52 Disc Harrow was a heavy duty trailed machine with a working width of 147in (3733mm) and weighing in at 2140lb (970kg), designed for use with the higher horsepower tractors in the 100 Series and to benefit from the use of their built-in Pressure Control System. It was equipped with two pneumatic wheels of 6.70 x 15 which functioned for depth setting and, when fully raised, for road transport. This was achieved by a double-acting ram powered by the tractor's hydraulic system. The two front gangs were each fitted with

The MF29 Seed Drill (20 row model) fitted with the optional 11-28 wheels: note the "reversed" tread pattern on the tyres to ensure a positive drive to the feed mechanism.

eight scalloped 20in (508mm) diameter discs spaced about 9¼in (235mm) apart, which gave good cutting and penetration. These were attached to the main steel fabricated frame at an angle of 20°. Each of the two rear gangs had ten plain discs of 20in (508mm) diameter set 7⅜in (188mm) apart, which could be mechanically set to 12°, 16° or 20°. Both front and rear gangs were equipped with adjustable scrapers, while their shafts ran on sealed self-aligning ball bearings. A central tine with a 6in (152mm) shovel was fitted to ensure that all the ground was cultivated in each pass.

At this time MF produced a range of different-width mounted disc harrows under the general heading of MF28.

In this next part we will consider the planting implements marketed by MF in the period covered by the UK-built 100 Series. (The MF726 Manual Potato Planter and the MF718 Automatic Potato Planter are described in my book *Massey Ferguson 35 & 65 Models in Detail.*)

New in this era was the MF29 trailed Grain and Fertilizer Drill, made in three different-width models, the 13-row which was 110in (2794mm) wide, a 15-row model at 125in (3115mm) wide and a 20-row machine at 162in (4114mm) wide. The two smaller models ran on 6.00 x 36 4-ply traction type tyres with the treads reversed; on the 20-row drill 6.00 x 36 6-ply tyres were fitted but 12.4/11 x 28 traction type could be specified at extra cost. The main frame and draw bar were fabricated from angle steel sections, while the hitch was of the clevis type. The seed and fertilizer hoppers were of mild steel, phosphate treated to resist corrosion. A full-width foot platform was provided at the rear to aid manual loading of the hoppers, and if it were deemed necessary a person could ride on the drill to keep an eye on its operation, although the driver of the tractor attached to the drill had a good view of the flow of both seed and fertilizer. MF offered a choice of three different types of openers – disc, Suffolk or hoe – as well as a choice of material, rubber or ribbon steel, for the conductors (the tubes taking the seed and fertilizer from the hoppers to the openers/ coulters). Depth control was by a hand-turned screw mechanism. The openers/coulters were staggered by 12½in (317mm), which enabled trash to pass between them more readily.

The main drive for this MF29 drill was taken from both wheels by roller chains to freewheel pawl boxes (similar to the freewheel on a rear bicycle hub) on the countershaft which provided differential effect when turning. The openers'/coulters' lift and feed cut-off were combined and effected mechanically by an eccentric sheaf and roller trip clutch operated by the

The Loader is an MF40 fitted to an MF135 which is loading an MF19 Manure Spreader operated by an MF165.

driver via a pull rope on the 13- and 15-row models. Only one such unit was fitted but the 20-row drill had two. The seed mechanism involved internal force feed by fluted rollers, with the actual seeding rate altered by a combination of an eight-speed oil bath gearbox (two on the 20-row model) and two alternative chain sprocket combinations. The fertilizer application rate was set by a 25-notch quadrant and hand lever controlling the shutter. This could be compounded by the interchanging of four chain sprockets. An acre meter was a standard fitment, calibrated to match the width of the machine. On the 20-row drill, disc markers were standard, while on the smaller models they were available as an extra but were in my opinion essential. A range of accessories was made available to widen the scope of these MF29 Drills, among them fertilizer placement coulters which kept the seed and fertilizer apart, a grass seed box, rubber closing flaps to cover some of the seed runs not required (14in, 21in or 28in/355mm, 533mm or 711mm rows were the requirement), wheel scrapers, and agitators to the seed hopper. Brush and wire type seed restrictors were available to help control the flow of small seeds, the type being matched to the flow characteristics of a particular seed.

During the period of 100 Series manufacture in the UK, MF marketed a semi-mounted drill, the MF34, which ran on 7.50 x 16 4-ply implement tyres and was available in 13- and 15-row form. The tractor's three-point linkage was used to lift the coulters out of work and to cut off seed and fertilizer flow. The seed feeding mechanism was of the fluted roller type, thus force fed. The fertilizer was fed by contra-rotating star wheels. A choice of two types of openers/coulters could be supplied, either discs or Suffolk type; these were assembled with a 12½in (317mm) stagger to improve trash flow. Also available was a facility for separate placement of seed and fertilizer. The 13-row model weighed 1520lb (690kg), and overall width was 9ft 5in (2.87m). The seed hopper had a capacity of 10cu. ft (283 litres) while the fertilizer hopper held 9.5cu. ft (269 litres). The 15-row model weighed 1708lb (775kg), and overall width was 10ft 7in (3.22m). Its seed hopper held 11cu.ft (311 litres), the fertilizer hopper having the same capacity.

Seeding rates per acre for both drills were quoted as 2lb to 7 bushels (1 bushel is equal to 8 gallons or 1.28cu.ft) or it could be doubled to 4lb to 14 bushels by using an alternative sprocket combination. Fertilizer rates per acre were 110lb (50kg) to 1200lb (545kg) of granular material, or reduced rates with different star wheels giving 56lb (25kg) to 110lb (50kg) per acre; according to MF these rates applied to both models.

In the realm of applying nutrients to the soil MF continued with the marketing of the MF712 Manure Spreader, a conventional wheel-driven machine, and the more developed PTO-driven MF19 Manure Spreader and MF22 Mounted Fertilizer Broadcaster, all three of which are fully described in my book *Massey Ferguson 35 & 65 Models in Detail.* The MF10 Spinner Broadcaster, introduced in 1966, was a linkage-mounted machine suitable for Category 1 and 2 attachment. It was PTO-driven, the hopper had a capacity of 1120lb (508kg) and the machine itself weighed 242lb (109kg). The MF12 Spinner Broadcaster, introduced the same year, had a larger capacity hopper holding 3360lb (1525kg). Being a trailed implement with a clevis drawbar and PTO-driven, it ran on a pair of 8.50 x 12 tyres and had an unladen weight 275lb (351kg). The quoted average spreading width was 25ft (7.62m) for granular fertilizer, for powder 17ft (5.18m). MF wisely did not quote application rates! Some of the special features of both machines were that the hoppers were painted with corrosion-resistant enamel and the open ends of the tubular frames were plugged to prevent internal rusting. These spreaders were capable of handling, in addition to granular fertilizer, lime, basic slag, grain salt and grit. The MF12 was fitted with a PTO-driven feed auger in the bottom of the hopper.

The MF32 Reciprocating Knife Mower of 1969 was in reality a developed version of the earlier MF732 Rear Mounted Mower introduced by MF into its

The MF32 mower. Note that some were painted grey.

range in 1962, which is fully described in my book *Massey Ferguson 36 & 65 Models In Detail.*

Interestingly, an MF Dealers publication of 1969 sets out in list form the numerous developments incorporated into the MF32 Mower over the earlier MF732 machine. It is worth noting the changes here in full:

- Cleaner, neater appearance
- A reduction in weight by 100lb (45.4kg) to 400lb (181.8kg) and in dimensions, allowing easier handling
- The swing frame is easily detached from the main frame
- Simple frame adjustment to suit either 52in (132mm) or 56in (1422mm) tractor wheel settings
- New break back latch gives improved performance hence better protection to the mower
- Full-width foot to parking stand gives better stability, facilitating attachment and detachment from the tractor
- Improved belt guard
- Category 1 and 2 linkage without changing pins
- Easier servicing, more accessible, and two fewer grease nipples (just 10. Big Deal!)
- Using the top link adjustment enables the cutter bar to be tilted from the tractor seat without tools
- More accessible belt adjustment and replacement
- Cutter bar lead adjusted by barrel nut
- More durable crankshaft bearing
- Dual purpose knife transport stay/knife puller
- Easier changing of knives – more clearance at cutter bar, hinge pitman can be fitted at any part of the stroke
- Fits a wide range of MF and competitive tractors

So reading through this long list of improvements, MF were obviously pretty proud of their development and achievement to the point that they went on to list the features retained from the earlier MF732 Mower, which are as follows:

- Employs the same British Standard cutter bar
- Adjustment of knife register by a threaded pitman (most mowers have fixed wooden pitmans so the knife register is adjusted by frame changes)
- Outer shoe design gives excellent crop separation and flow
- Inner and outer shoes shimmed to give true ledger alignment, so efficient cutting extends to the extremities of the cutter bar
- Adjustable locking lift link safeguards knife head and pitman from the effects of violent bouncing of the cutter bay
- Well proven knife head and pitman
- Long swath board gives clear lane for tractor wheels
- Crop divider available as an accessory for use in laid tangled crops, which ensures a clear lane for the next round
- Compact storage
- Inner swath board available as an accessory, leaving narrower swath – useful when following up with a forager.

The MF51 Rotary Mower.

The reciprocating mower days were numbered when rotary grass cutters for agricultural use were introduced. MF's first offering was the 51 Rotary Mower of late 1969. It was a design aimed to utilise the power output of the larger MF tractors in the 100 Series (25hp was quoted as the minimum PTO requirement) but they could deal efficiently with much more and at a greater speed. Rotary mowers were also able to work very efficiently with laid grass and generally speed up the whole mowing operation, particularly through not having to be constantly re-sharpening the many knife sections of a conventional reciprocating mower.

The working principle of the Rotary Mower is a simple one. It cuts by impact, not by a shearing action. Its elements comprise a cutter bar 65in (165mm) wide equipped with four horizontal revolving discs each with two protruding double-edged knives which could be interchanged and easily re-sharpened. Each of these assemblies revolved at high speed (3500rpm, requiring a PTO speed of 600-640rpm) just above the ground. The two inner discs rotated in a clockwise direction while the two outer discs turned anti-clockwise. This feature tended to bring the crop into a narrow, well fluffed swath. Drive from the tractor's PTO is taken through the normal universally-jointed telescopic shaft with an over-run clutch at the machine end. From there, within a sheet metal casing, the drive is transmitted by four "V" belts to a step-up ratio bevel gearbox and from that via gears inside the cutter bar to the high speed discs. The outside disc carries a rotating drum that performs the function of a swath board or divider; this together with an adjustable deflector forms the crop, a freely aerated swath, thus creating a clean lane for the mower's inner shoe on the next round. The whole cutter bar is raised manually by a cranked handle into the vertical transport position; to ease this procedure a powerful tension spring is incorporated into the lifting linkage.

The great benefit of this type of mower over the earlier reciprocating knife type was that they rarely, if ever, blocked when operating in tangled crops, while the fact that the replaceable knives were attached to the rotating discs by only one bolt meant that if they were to hit a small obstruction they would swing back. An essential safety feature was the provision of a break-back device to the cutter bar, thus avoiding damage if it were to strike an obstruction. Another important safety feature was the provision of a heavy fabric cover over the whole of the cutter top and side to ensure that stones were not flung far and wide at high speed but were contained within this cover. As an optional accessory, extra-height skids were available to enable the mower to be used for pasture topping.

Next comes the MF711 Potato Harvester, which was the result of collaboration between N.I.A.E Silsoe (National Institute of Agricultural Engineering) and MF development engineers, the aim being to produce a machine that took the backbreaking effort out of harvesting potatoes or bulbs by hand.

This harvester was a trailed machine running on two 6.50 x 16 ribbed tyres. The height of each in relation to the frame of the harvester could be adjusted by two double-acting hydraulic rams controlled by the tractor driver so as to keep the machine on a level plane whilst traversing sideling ground. This harvester was PTO-operated, with each of the drives to the various components being taken from the machine's main gearbox and each output being protected by individual slip clutches. No chains or belts were fitted to the basic bagger harvester, but if the machine was built or converted with an elevator to discharge the crop into a trailer moving alongside both belts and chains were incorporated into it.

MF claimed that even the small MF130 tractor could handle the harvester. The instruction book not only gives details of setting up the machine for MF tractors but also its attachment to the Fordson Super Major and Dexta as well as the Nuffield 3 and 4 tractors.

In addition to the tractor driver three or generally four pickers would be needed, and if the harvester was a bagger model an operator would also be needed to deal with that aspect. If on the other hand it was equipped with an elevator discharging into a trailer running alongside, obviously a driver was required for that. It should be noted that the basic bagger model could be quite readily converted with the addition of an elevator unit.

The principal features of how the MF711 Potato Harvester worked are as follows. Firstly the haulm must be removed about a fortnight before the commencement of harvesting, which could be done by chemical spray or better still by using the MF760 Forager. One row of potatoes is harvested in each pass, with the tractor straddling two rows and the harvester's actual digging/lifting rotor tines in line with a third row to the right-hand side of the tractor. MF recommended that the famous Ferguson Steerage fin that was part of the harvester's equipment be deployed to aid accurate operation.

The main lifting element of this machine is a slightly inclined, slow-turning rotor made up of 27 curved steel tines, each of ½in (12mm) diameter. For lifting small potatoes or bulbs these could be interspersed with a similar number of slightly smaller-diameter tines curved to the same profile. To the left of this rotor is a concave steel rotating disc with an adjustable scraper, which was set to lightly cut a line in the soil to assist the effective operation of the lifting rotor. To the right of this was a 5ft (1524mm) diameter by 10in (254mm) wide open slow-turning "drum" (see illustration) made up of seven drum wires; again these could be interspersed with eight additional wires when harvesting small potatoes or bulbs. On the inside periphery of this open drum, across its width, at intervals were eight cross bars each equipped with four short spikes of round steel bar capped off with rubber fingers. The first row nearest to the digger rotor was relatively short, the next ring of rubber fingers were slightly longer, while the two outermost rings were again slightly longer. The digger rotor eases the potatoes out of the soil and at the same time moves them towards and into the open slow-turning drum. This action in essence not only lifts the crop to deposit it on to the sorting table, which turns at seven revolutions per minute, but also at the same time releases loose soil from the tubers. To keep the drum wires clear of trash, particularly in damp field conditions, the MF711 Harvester was supplied with an adjustable drum cleaner made up of eight rubber-coated finger wheels sandwiched together and mounted on a free-turning spindle; these were turned as the drum rotated by touching the cross bars of the lifting drum. Generally when set and in use the fingers were set to protrude about 1½in (38mm) into the drum. In dry conditions this facility was not needed so it would be withdrawn to save wear.

The sorting table consisted of an upper deck on to which the potatoes were deposited, to be picked over by the team of three or four pickers (depending on the crop level) and then placed on the lower rotating deck where they went on to be discharged into the bagging-off chute, which required one operator to deal with that aspect of the harvesting operation. The remaining trash and stones went on to be deposited back onto the field. As an alternative to the bagger,

This view taken from an early MF sales brochure gives a good idea of the layout of the MF711 Potato Harvester from the point of view of the operators.

MF offered as an accessory a mechanically driven side elevator, the angle and reach of which could be altered hydraulically by the tractor driver. Thus the potatoes were offloaded into a tipping trailer moving alongside.

It should be noted that the drive for the primary elevator is taken via a jointed Cardan shaft from the main gearbox of the harvester to a vee pulley close to the bottom of the elevator, then via a long vee belt to the roller bars on the top end of the primary elevator. This drive not only powers the primary elevator's flat rubber conveyor belt but also turns a roller chain sprocket that feeds motion to the secondary elevator's conveyor belt. The use of a chain drive between primary and secondary elevator prevents the possibility of crop damage: if for some reason a blockage occurs on the secondary elevator vee belt, slippage takes place and both elevators stop running.

It can be appreciated that with all this gentle movement of the crop over slatted surfaces excellent separation of potatoes and soil takes place, but it is ultimately the diligence of the pickers that ensures all the good potatoes are taken off.

The MF20 3-ton low loading trailer; each of the sacks would weigh 1cwt (50.8kg).

It seems appropriate here to give a brief account of the specification of this interesting MF design. Height (lowered on rams) is 5ft 5in (1651mm), height raised is 6ft 3 n (1905 mm), or for the bagger model with elevator in transport position it is 9ft 4in (2845mm). The width with side operator platform lowered is 8ft 4in (2540mm) and with side platform raised it is 9ft 1in (2768mm). With the elevator in the transport position the machine is 9ft (2743mm) wide. Length, including draw bar, is 13ft 1in (3987mm). The weight of harvester is only 2688lb (1219kg), or with the elevator attached it is 3024lb (1371kg).

To round off this section on the MF711 Potato Harvester it seems worthwhile to mention that I have, in my archives, a comparative study done in the mid-sixties in which the MF711 operated alongside competitors' machines. In my opinion it fared very well, particularly in the area of low damage to crop.

Coming now to the last part of this lengthy chapter we will consider the range of trailers and the one basic loader model MF marketed during the period covered by this book.

MF continued with the sale of MF17 3-ton tipping trailer and the 5-ton MF18 but only until early 1969, when a new, much wider range of six trailers was introduced, the carrying capacity of the basic trailers being 3 tons to 7 tons. These trailers were manufactured by Weeks under a sub-contract and were generally of all-steel construction, thus rather prone to rusting due to the extensive use of cheap cold-formed pressed steel sections.

We will first consider the MF20, a 3-ton low loader and non-tipping trailer. The wheels were shod with 7.50 x 10 10-ply tyres, with the axle rearward of the centre line of the body and tucked under the side members of the trailer bed, which was 174in (4.4m) long, 72in (1.8 m) wide and only 24in (609mm) from the ground. The draw bar coupling was an interchangeable ring or clevis, and a built-in jack was included as was a parking handbrake. Two fixing positions were provided for the axle attachment bolts. The trailer in standard form weighed 1470lb (668kg). Available as accessories were hay lades, a head board and a road lighting set.

The MF21 Tipping Trailer was designed as a tough general purpose unit with a rated carrying capacity of 3½ tons. The load platform was 120in (3.0 m) long and 72in (1.82m) wide. The standard tyre size was 7.50 x 16 8-ply but at extra cost either 9.00 x 16 or 10.00 x 16 tyres could be supplied. The metal sides were 17in (43mm) high, each of one piece and could

The smallest Tipping Trailer, the MF21.

The High-Level version of the MF22 Trailer.

be hinged down. The head board was fixed, while the trail board was top-hinged and equipped with a quick-release latch so it could be readily taken off. The body tipped hydraulically to an angle of 56°. The unladen weight of the basic steel trailer was 1680lb (763.6kg). Offered as extras were a hardwood floor, steel extension sides for grain, a sacking-off chute on the tail board, a rear platform, silage extension sides, hay lades and a road lighting kit. It was a very popular trailer in its day.

The MF22 tipping trailer, again manufactured by Weeks, was basically similar to the MF21 but was built bigger and stronger to enable it to be rated at 4½ tons. With the standard 17in (431mm) dropdown sides it had a capacity of almost 4 cu.yds. The body was 130in (3.3m) long by 78in (1.98m) wide and weighed in standard form 1792lb (814kg). Three different axle location points were provided and as one would expect the hitch had interchangeable ring or clevis attachment, with a built-in screw jack. The angle of tip was 51°. The accessories made to match this trailer were similar to those offered for the MF21 but did not include a hardwood floor.

The MF23 Tipping Trailer fitted with silage sides.

The next trailer in the numerical sequence is the MF23, also produced by Weeks. It was known as a High Level Tipper and rated to carry 4½ tons. The body was of the same dimensions as the MF22 but the tyre size was increased to 12.00 x 16 8-ply to give better stability when tipping at high level. The high lift was achieved by having a scissor lift frame operated by its own dedicated hydraulic ram directly above the main chassis of the trailer. The top frame of the scissor lift supported the body, which was tipped 51° in the normal way by a second hydraulic ram. To enable the tractor driver to easily control the movement of each individual ram a diverter valve was mounted on the trailer's hand brake bracket to enable precise control of each ram under hydraulic pressure. The accessories were the same as the MF22. This type of trailer was popular with some users in its day before the advent and widespread use of telescopic handlers on farms.

The MF24 was a tough all-steel tipping trailer rated at 6-ton load with a capacity of 6 cu.yd. The bed was 147in (3.73m) long by 84in (2.3m) wide, with 24in (609mm) sides that hinged down and were detachable. The draw bar was ring hitch only. The tyre size was 12.00 x 16 10-ply and the heavy axle had the option of four locating points depending on the amount of weight transfer required. Accessories for this model were grain sides, sacking-off chute, silage sides and lighting kit.

Largest in the trailer line-up was the MF25, rated to carry 7 tons. These all steel trailers had fixed sides of 42in (1.02m) giving a bulk capacity of just over 11 cu.yd. They were equipped with twin axles mounted on patented rubber suspension units which together with four 10.00 x 16 8-ply ribbed tyres gave very good flotation. Each wheel was fitted with a 9in (228mm) diameter drum brake controlled by a lever within easy reach of the driver; it also doubled as a parking brake. Twin two-stage hydraulic rams ensured that the body tipped readily to 51°. The draw bar terminated with a 2in (50mm) pick-up ring. Accessories available were silage sides (factory fitted), a sacking-off chute and a lighting set. It is this type of trailer that evolved into the 15-ton-plus trailers so commonly used on farms and highways today.

In September 1970 MF introduced the MF40 loader, which could be fitted to all the tractors in the 100 Series and was produced at Banner Lane. The loader was designed in such a way that it was compatible with tractors with or without a cab. As the main lifting frame of the loader was interchangeable across the range, the variation in tractor size was taken care of by the design of the brackets. The MF135 and MF148, having a shorter engine than the four-cylinder models, required a heavy plate to be fitted to the underneath of the front axle support casting. On the larger models two independent brackets were bolted to tapped holes provided either side of the front axle support casting. A number of features were incorporated into the loader layout to facilitate fitment and detachment. The pivot posts mounted

The MF24 6-ton Tipping Trailer off to collect another load of grain.

either side of the tractor could remain in place and not interfere with rear-mounted implements; these side members incorporated a neat stowage arrangement for the hydraulic rams. The parking stand provided made it simpler to remove and store the loader, and also acted as a transport lock to ensure safe high-speed transport.

To team up with this front loader MF marketed six different attachments to widen the scope of the basic machine. The manure fork was 43in (1.09m) wide with eight square chisel tines, an average lift capacity of 6cwt (305kg), and a dirt plate to convert the fork to a scoop. The dirt bucket had a hydraulic or mechanical trip, was 42½in (1.08m) wide and had a capacity of 10cu.ft. The bulk materials bucket had a mechanical trip, was 78in (1.9m) wide and had a capacity of 20cu.ft. The root crop bucket was 64½in (1.64m) wide with nineteen 10in (254mm) long tines; its capacity was 17.25cu.ft. The pair of forklift tines was capable of lifting up to 10cwt (508kg) and had a parallel action to keep the forks horizontal throughout the lift cycle. The dozer blade was 66in (1.68m) wide and was useful for backfilling and grading.

Alongside these attachments MF also made available a range of front grille guards to suit individual models and also a heavy counterweight for the rear three-point linkage. Also offered was a selector valve plumbed into the tractor's hydraulics which enabled the three-point linkage to be isolated from the lift rams.

Just to round off this rather lengthy chapter it seems fair to say that the era of the 100 Series tractors ushered in a trend that is very much with us today – the need to achieve a greater output of food per man-hour invested.

The MF25 dual axle Tipping Trailer with silage sides.

MF168 Super-Spec fitted with an MF40 loader filling an MF19 Manure Spreader (the 130-bushel version).

Chapter 11

Conversions & Accessories

The MF 100 Series tractors produced at Banner Lane were considered to be pretty well equipped, even in the basic form, and this was to improve with the introduction of the early rather utilitarian safety cabs. With the introduction of Q cabs the driver's environment became almost luxurious!

MF introduced the 500 Series, which replaced the 100 Series, at the Royal Show in 1975, but it was not until 1978 that they felt it appropriate to launch their own range of tractors built with 4WD alongside the 2WD models. Prior to this and within the 100 Series production span at Coventry, customers wishing to gain the benefits of 4WD had to turn to outside manufacturers who offered conversion kits. Before proceeding with descriptions of these after-sales fitments it seems worthwhile to recount briefly the early development of 4WD systems applicable to Ferguson and MF tractors.

The front cover of a Four Wheel Traction sales brochure relating to the MF165.

Early in the days of the TE20 Dr Sion Segre Amar, together with a short-term partner, Carlo Torchio, established a business in Nichelino, near Turin in Northern Italy to import second-hand tractors from England (mainly Fordsons and Fergusons). The tractors were then refurbished and often fitted with new Perkins diesel engines, for which Dr Segre Amar was the main Italian agent. It has been suggested that he chose to name his firm Selene, after his wife, and he later chose the name of his son Manuel for the 4WD kits which he designed and sold.

Very early in his venture Dr Segre Amar developed a 4WD system for Fordsons and Fergusons as well as certain continental makes. The first developed was known as System I, and eventually a further four systems were developed. The difference lay in how the drive for the front axle was picked up from the existing transmission. System I, used on Fordson Majors and Ferguson TE20s, consisted of a sandwich transfer box installed between the rear of the gearbox and the back axle housing. Within this transfer box was a sliding gear shifted by a hand lever which enabled two- or four-wheel drive to be selected. The front axles were sourced from ex-military vehicles, Jeep for the TE20s and GMC for Fordsons and MF35s and 65s.

Another player, Major Phillip Henry Johnson, the managing director and chairman of Roadless Traction, had been involved in the long-term development of track-laying systems for a wide range of applications. By 1952 Johnson had met up with Segre Amar while in

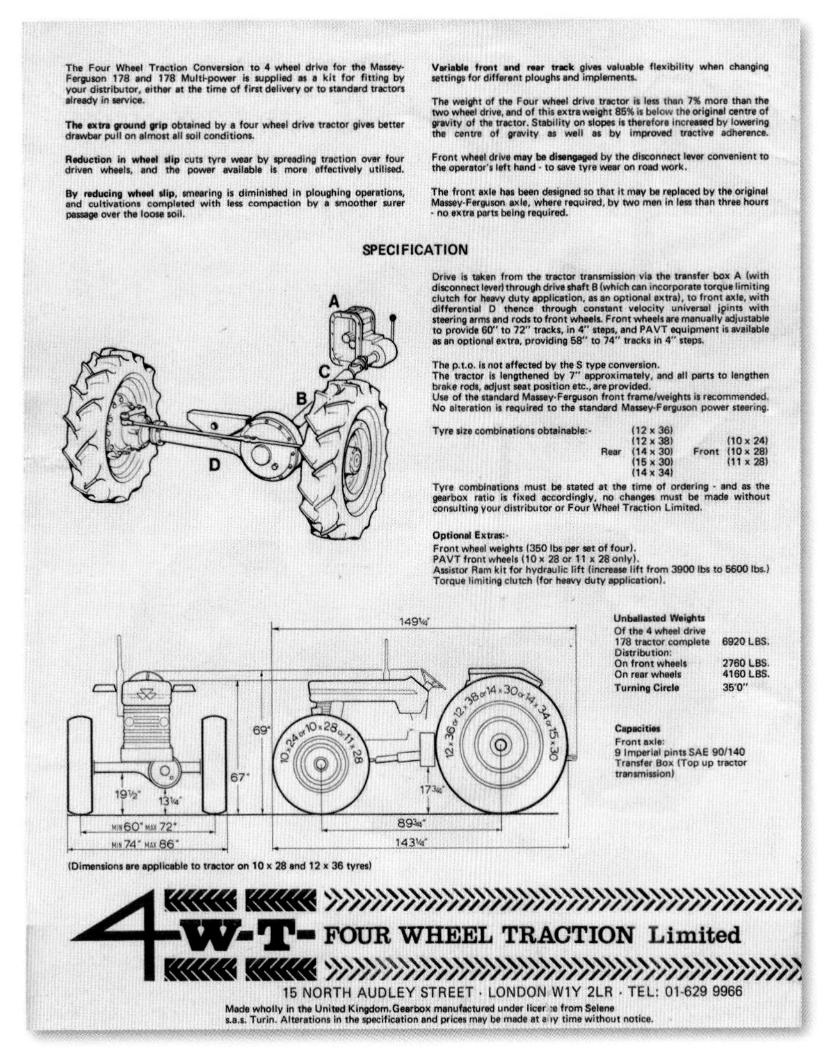

The Four Wheel Traction Conversion to 4 wheel drive for the Massey-Ferguson 178 and 178 Multi-power is supplied as a kit for fitting by your distributor, either at the time of first delivery or to standard tractors already in service.

The extra ground grip obtained by a four wheel drive tractor gives better drawbar pull on almost all soil conditions.

Reduction in wheel slip cuts tyre wear by spreading traction over four driven wheels, and the power available is more effectively utilised.

By reducing wheel slip, smearing is diminished in ploughing operations, and cultivations completed with less compaction by a smoother surer passage over the loose soil.

Variable front and rear track gives valuable flexibility when changing settings for different ploughs and implements.

The weight of the Four wheel drive tractor is less than 7% more than the two wheel drive, and of this extra weight 85% is below the original centre of gravity of the tractor. Stability on slopes is therefore increased by lowering the centre of gravity as well as by improved tractive adherence.

Front wheel drive **may be disengaged** by the disconnect lever convenient to the operator's left hand - to save tyre wear on road work.

The front axle has been designed so that it may be replaced by the original Massey-Ferguson axle, where required, by two men in less than three hours - no extra parts being required.

SPECIFICATION

Drive is taken from the tractor transmission via the transfer box A (with disconnect lever) through drive shaft B (which can incorporate torque limiting clutch for heavy duty application, as an optional extra), to front axle, with differential D thence through constant velocity universal joints with steering arms and rods to front wheels. Front wheels are manually adjustable to provide 60" to 72" tracks, in 4" steps, and PAVT equipment is available as an optional extra, providing 58" to 74" tracks in 4" steps.

The p.t.o. is not affected by the S type conversion.
The tractor is lengthened by 7" approximately, and all parts to lengthen brake rods, adjust seat position etc., are provided.
Use of the standard Massey-Ferguson front frame/weights is recommended.
No alteration is required to the standard Massey-Ferguson power steering.

Tyre size combinations obtainable:-

	Rear		Front
	(12 x 36)		
	(12 x 38)		(10 x 24)
Rear	(14 x 30)	Front	(10 x 28)
	(15 x 30)		(11 x 28)
	(14 x 34)		

Tyre combinations must be stated at the time of ordering - and as the gearbox ratio is fixed accordingly, no changes must be made without consulting your distributor or Four Wheel Traction Limited.

Optional Extras:-
Front wheel weights (350 lbs per set of four).
PAVT front wheels (10 x 28 or 11 x 28 only).
Assistor Ram kit for hydraulic lift (increase lift from 3900 lbs to 5600 lbs.)
Torque limiting clutch (for heavy duty application).

Unballasted Weights
Of the 4 wheel drive
178 tractor complete 6920 LBS.
Distribution:
On front wheels 2760 LBS.
On rear wheels 4160 LBS.
Turning Circle 35'0"

Capacities
Front axle:
9 Imperial pints SAE 90/140
Transfer Box (Top up tractor transmission)

(Dimensions are applicable to tractor on 10 x 28 and 12 x 36 tyres)

4-W-T- FOUR WHEEL TRACTION Limited
15 NORTH AUDLEY STREET · LONDON W1Y 2LR · TEL: 01-629 9966
Made wholly in the United Kingdom. Gearbox manufactured under licence from Selene s.a.s. Turin. Alterations in the specification and prices may be made at any time without notice.

The back cover of a similar brochure but this time referring to the MF178 tractor.

Italy and eventually an agreement was reached between the two firms. Roadless Traction would manufacture Selene Manual conversions under licence but only for Fordson tractors. All of Selene's five systems were protected by patents. System IV was developed to allow 4WD conversions to be fitted to certain Landini and Fiat tractors, as well as the MF35/65, 130, 135, 165 and 175. This System IV exploited the fact that these MF tractors all had the facility to drive the PTO relative to ground speed. Dr Segre Amar found that by fitting a transfer box with a built-in selector directly to the tractor's PTO a cost saving could be made as the tractor did not have to be split to install the sandwich transfer box. The only drawback to this arrangement was that the PTO could not be used if 4WD was needed; an extension of the tractor's PTO passed through the casing so that limited use of the PTO could be made.

By 1948 Robert Eden had established a business in Kent to buy second-hand tractors from around Britain and then export them to other countries. A high proportion of them found their way to Selene for refurbishment and were often converted to 4WD. William Fuller joined Robert Eden's company in 1950 and two years later bought Eden out.. Fuller felt it worthwhile to offer imported Selene Manuel 4WD conversion kits to his farming customers with MF tractors, and by the late 1950s he had fitted Selene conversions to MF35s and 65s. In due course he became concerned that the Selene PTO transfer box was not robust enough, so he produced a strengthened version. Although outwardly very similar the two boxes can be readily identified: the early type had Selene cast into the rear cover plate while the later stronger type had RE cast into it.

The drawback of System IV in not being able to use 4WD if PTO was required was overcome in 1965 when Fuller was able to reach an agreement with Segre Amar allowing RE (Robert Eden) to manufacture a sandwich-type transfer system, thus getting around the drawbacks of System IV. This became the foundation stone of a new business set up by Fuller and known as Four Wheel Traction, with its head office at 15 North Audley Street, London.

I'm sorry for this rather long lead-in to these early forays into the agricultural application of 4WD but I felt that it was necessary to set the scene before embarking on a description of how some of these systems were applied to MF tractors within the 100 Series built at Coventry. Although Dr Segre Amar of Italy sold kits directly to his countrymen, in the UK it was Four Wheel Traction who not only supplied the UK market but exported as well. William Fuller's earlier 4WD systems of taking drive from the PTO were rather short lived because of their inherent disadvantages.

Reproduced here are the front cover of the Four Wheel Traction sales brochure for the MF165 (left) and the reverse side of another brochure for the MF178 (above), which should give readers a full insight into the layout and specification.

Less well known is the fact that Four Wheel Traction designed and manufactured for fitment to MF100 Series tractors two types of sandwich reduction gearboxes. These were deemed necessary by some customers needing to operate implements at very low speeds, for instance planters, rotary cultivators, hedge cutters, etc., but they were not suitable for exceptionally heavy drawbar work.

The simplest one gave a reduction of 4:1 when engaged. The main element of this assembly was a standard MF epicyclic gear unit used to give low ratio selection within the standard MF transmission. These units were installed behind the main gearbox and forward of the final drive casing. Their fitment in no way

Bernard Curtis' well-used MF168 with four-wheel-drive conversion by Four Wheel Traction.

The front axle of the 4WT MF168.

interfered with the operation of the PTO but it did increase the wheelbase of the MF135, 165, 175 and 185 tractors, so part of the kit for these models would include extensions for the brake linkages and foot boards.

When these conversions and those for the 4WD were ordered new from the MF dealer, the standard cab for that model would not be fitted at the factory and a longer cab would be fitted by the dealer following

4-W-T-

REDUCTION GEAR BOX
(4:1)

For Massey-Ferguson tractors 35, 65, 135, 148, 165 and 168
140, 145, 150 and 155 (Export).

The 4:1 reduction gearbox slows the speed of the tractor (in low range) to approximately ¼ that of the transmission.

It can be disengaged with a selector, allowing the option of normal, or crawl, speeds.

It provides double the number of speeds obtainable in the LOW range (with or without Multipower). Speeds in the HIGH range are not affected.

6 Speed Standard Transmission + Reduction Gearbox
= 9 speeds forward 3 reverse

8 Speed Standard Transmission + Reduction Gearbox
= 12 Speeds forward 3 reverse

Multipower Transmission + Reduction Gearbox
= 18 speeds forward 6 reverse

(Independent or live) P.T.O. speeds in all gear positions, are maintained at the same levels as on the standard tractor. Ground speed P.T.O. remains proportional to the forward speed of the tractor.

Massey-Ferguson safety cabs are suitable without alteration for 148 and 168 - when fitted with Reducton Gearbox. 135 tractors should be ordered with 148 cabs. 165 tractors will require new 168 cabs.

Provides slow tractor speeds, with normal P.T.O. operation - for planting, rotary cultivating, seedbed preparation, hedging, drainage work and all slow speed operations.

Available from your Massey-Ferguson Distributor.

Manufactured by

4-W-T- FOUR WHEEL TRACTION Limited

15 NORTH AUDLEY STREET · LONDON W1Y 2LR · TEL: 01-629 9966

CABLES. FORWETRAC. LONDON WIY 2LR ENGLAND

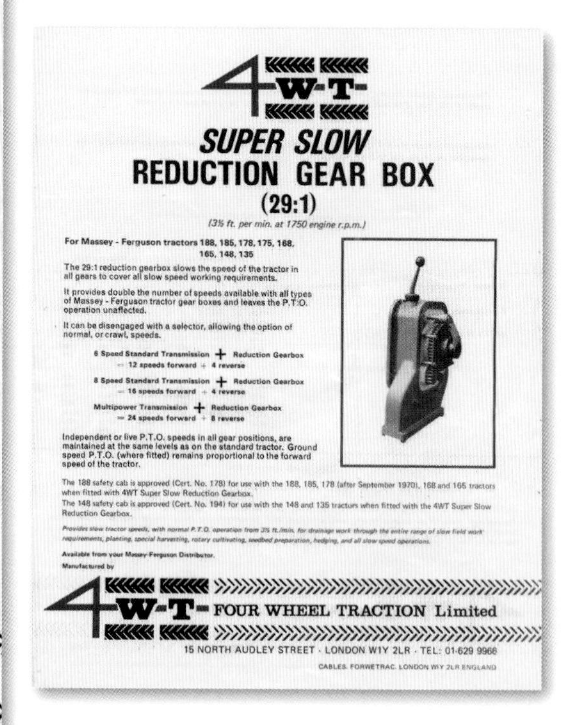

4-W-T-

SUPER SLOW
REDUCTION GEAR BOX
(29:1)

(3½ ft. per min. at 1750 engine r.p.m.)

For Massey - Ferguson tractors 188, 185, 178, 175, 168, 165, 148, 135

The 29:1 reduction gearbox slows the speed of the tractor in all gears to cover all slow speed working requirements.

It provides double the number of speeds available with all types of Massey - Ferguson tractor gear boxes and leaves the P.T.O. operation unaffected.

It can be disengaged with a selector, allowing the option of normal, or crawl, speeds.

6 Speed Standard Transmission + Reduction Gearbox
= 12 speeds forward + 4 reverse

8 Speed Standard Transmission + Reduction Gearbox
= 16 speeds forward + 4 reverse

Multipower Transmission + Reduction Gearbox
= 24 speeds forward + 8 reverse

Independent or live P.T.O. speeds in all gear positions, are maintained at the same levels as on the standard tractor. Ground speed P.T.O. (where fitted) remains proportional to the forward speed of the tractor.

The 188 safety cab is approved (Cert. No. 178) for use with the 188, 185, 178 (after September 1970), 168 and 165 tractors when fitted with 4WT Super Slow Reduction Gearbox.
The 148 safety cab is approved (Cert. No. 194) for use with the 148 and 135 tractors when fitted with the 4WT Super Slow Reduction Gearbox.

Provides slow tractor speeds, with normal P.T.O. operation from 3½ ft./min. for drainage work through the entire range of slow field work requirements, planting, special harvesting, rotary cultivating, seedbed preparation, hedging, and all slow speed operations.

Available from your Massey-Ferguson Distributor.

Manufactured by

4-W-T- FOUR WHEEL TRACTION Limited

15 NORTH AUDLEY STREET · LONDON W1Y 2LR · TEL: 01-629 9966

CABLES. FORWETRAC. LONDON WIY 2LR ENGLAND

A basic reduction gearbox from Four Wheel Traction (left) and the more sophisticated Super Slow 'box (above).

The Coldridge Collection's MF135 with 4WD conversion by Schindler.

the installation of either 4WD or reduction gearbox. In the case of the MF Super-Spec tractors none of this was an issue because the sandwich gear of the transfer box simply replaced the spacer.

Later, Four Wheel Traction went on to design and produce a super-slow reduction gearbox with a ratio of 29:1. It was often fitted to 100 Series tractors modified to operate trench-digging machines at very slow forward speeds. These units were 6in (150mm) long but with a much lower profile to the casing to accommodate the necessary number of spur gears. To give an idea of the reduction available for an MF188 with an eight-speed transmission with the Super Box engaged, the slowest forward speed with the engine running at 1500rpm is 3.06ft/min (932mm/min) while with the normal transmission the slowest forward speed in gear one at the same engine rpm is 92.14ft/min (28.0m/min). It should be noted that both of these reduction boxes had their gear selector lever mounted on the top of the casing, as are the normal gear levers. Later, when MF went on to produce their own design of reduction box, the selector lever was positioned on the left-hand side of the casing..

The Coldridge Collection did at one time have on loan from the Fuller family an MF35X tractor fitted with both 4WD and 4:1 reduction units, making the wheelbase 12in (305mm) longer – very suitable for tall tractor drivers!

A further 4WD conversion was made by Schindler of Switzerland, who took over the Selene business in 1970. In their layout, drive to the front axle is taken from the tractor's transmission by the installation of a special casting incorporating a built-in selector lever on the left-hand side which replaces the plate that normally locates the PTO selector lever. Needless to say this casting has two levers protruding from it, one for PTO and one for 4WD. Drive is then taken by cardan shaft running on the left of the engine via a torque-limiting clutch to the front axle differential – all in all a very neat and well engineered conversion which was promoted by German and Austrian MF Dealers.

An MF135 fitted with the Schindler 4WD in The Coldridge Collection (imported from Austria) is also fitted with a mid-mounted PTO. Although built in Coventry, models exported to Germany and Austria were fitted with a different transmission housing that incorporated a facility to fit a mid-mounted PTO if it was needed; this again had its own selector lever on the right-hand side. Mid-mounted mowers were popular in these countries.

To widen the scope of tractor sales MF canvassed outside manufacturers of specialised machines to buy their tractors or skid units to team up with their own equipment. For example, Standen Engineering of Ely Cambridge used the MF135 to power their Solo MkII sugar beet harvester, and in 1970 as an optional fitment to the Cyclone beet harvester.

Shawnee Poole of Cardiff, who manufactured dumper

Front hub of the Schindler-converted MF135.

A torque-limiting clutch protects the front axle.

trucks and construction equipment, used the MF178 tractor to haul their 12-ton swan-neck rear dump trailer. These tractors were fitted with an air braking system that operated the brakes on the trailer, and for highway use Shawnee Poole offered a braking system that fully complied with The Road Traffic Act. The tipping of the 12 ton dumper was effected hydraulically by two rams that gave a tip angle of 60° in approximately ten seconds. The power for this operation was supplied by an engine-mounted hydraulic pump capable of delivering oil at 2400psi (168kg/cm^2) with an output of 11.5gal/min (52 litres/min). The oil reservoir mounted on the right-hand side of the engine just forward of the cab had a capacity of 5gal (22.7 litres). Shawnee Poole also offered customers the option of a hydraulic drive to the dumper's wheels.

The lever on the right engages four wheel drive. The inner lever engages the side PTO.

Another item of industrial equipment MF marketed was a Bristol Duplex 2 Stage Compressor, engineered to fit, with appropriate brackets, to the industrial MF20, MF40 and MF50. The compressor was resiliently mounted on its frame and PTO-driven via a supplementary shaft and six B-section vee belts. The compressor had an output of 140cu.ft/min (3.96cu.m/min), British Standard 726, at a working pressure of 100psi (7kg/cm^2) so it could run two heavy duty road breakers. MF tractors thus equipped were very often fitted with a front loader, thus greatly increasing their versatility.

The other firm that used MF parts, as the basis of some rough-terrain forklifts, was Sambron of High Wycombe, Buckinghamshire. Their three 2WD machines had lifting capacities of 2 tons, 2½ tons and 3 tons, with a range of attachments to widen the machines' versatility. All three models were powered by the Perkins AD3-152 diesel engine, rated to produce 46.5bhp at 2200rpm. This was connected to an MF165

The Schindler-converted MF135's is fitted with a mid-mounted mower driven by a mid-mounted PTO.

four-speed transmission with a high and low range as well as a reversing unit, thus giving eight speeds in either direction. The brakes were of the enclosed disc type. The rear wheels were steered by a hydrostatic system.

Conversions for some of the 100 Series tractors (mainly MF165s) were produced by County Commercial Cars of Fleet in Hampshire. Although their name is generally associated with Ford tractors and commercial vehicles they also carried out work for other manufacturers including MF, which was a world leader in the mechanisation of sugar cane production, especially in Australia, South Africa and the Caribbean. Following market research MF established the need for a very high clearance tractor and associated equipment. This resulted in the conversion firstly of the MF65 MkII and later the MF165 by County to produce the Hi Hi, which had a full underbelly clearance of 2ft 9in (838mm). The Hi Arch tool carriers were designed by

This tractor came from Austria. German and Austrian tractors are often fitted with this type of height-adjustable hitch.

A sales brochure distributed by the German Massey Ferguson Dealer Gebrüder Schoeller of the Schindler four-wheel-drive conversion. The driver with his sleeves rolled up must be a tough chap!

MF to match the converted 165 but were manufactured in the UK by Bomford and Evershed at Evesham and in the USA by MF Fowler based in California. These Hi Hi 165s ran on 7.50 x 20 front tyres with extended kingpins braced by tubular steel members which were telescopic to allow for track width adjustment. The rear tyres were 12.00 x 30. To achieve the underbelly clearance, drop gearboxes were developed incorporating a track of spur gears to take power from the tractor's differential output shafts to the wheel hubs, where the epicyclic reduction gears were fitted. This was designed to ensure that the train of spur gears did not have to transmit excessive torque loadings and the arrangement in effect gave a strong portal rear axle.

Mention must also be made of the MF165 Semi Hi Clearance model, which was often used in forestry work as the slightly raised clearance reduced the possibility of damage caused by striking tree stumps. The standard 165 had a ground clearance of 13.7in (350mm) while the Semi Hi had a clearance of 15.7in (400mm). This increase in height was achieved by the fitting of 6.00 x 19 front tyres and 12 x 36 rears. The Normal High Clearance 165 had a ground clearance of 18.9in (480mm) with extended kingpins. It had 6.00 x 16 front tyres and the rear wheels were fitted with 12 x 38 tyres. The transmission ratio was changed to ensure that the road speed remained more or less the same as a standard MF165s.

To round off this chapter it seems prudent to list some of the accessories that MF marketed during the era of the 100 Series to enhance the range of applications they were able to perform.

A Dual Spool and Selector Valve Kit was produced that was mounted on the top of the rear transmission housing. This addition provided hydraulic control over two remote single-acting hydraulic cylinders. The selector valve provided a means of operating single-acting rams from the tractor's control quadrant. Thus it could be used in conjunction with the dual spool valve to provide a means of controlling three single-acting cylinders.

A Two Spool Valve and Combining Valve, mounted on the left-hand footrest, could be used to control two single- or double-acting remote hydraulic rams. The combining valve provided a means of uniting the flow of the tractor's hydraulic lift pump and that of a high-capacity auxiliary pump. This ensured faster response when operating remote hydraulic cylinders.

For added driver comfort MF offered, at extra cost, a spring suspension seat that could be adjusted to suit the weight of the operator and the operating conditions. This type of seat also featured fore and aft adjustment as well as height selection.

MF made available a kit of parts to modify the Category 1 linkage of the MF135 so that both Category 1 and 2 implements could be mounted. Basically the kit

Self-propelled
Sugar beet harvester
Standen
Cyclone MkIV

Standen's MF135-based Cyclone MkIV self-propelled beet harvester.

An MF135 powering a Standen Solobeet model.

Shawnee Poole offered these MF178 rear dump haulers.

The Bristol Duplex Compressor mounted on a 100 Series tractor.

consisted of a pair of reversible lower links, i.e. Category 1 at one end and Category 2 at the other, with two bushes to be fitted at the tractor end when a Category 1 implement was hitched. A modified top link was also included.

An Automatic Pick Up Hitch was also available, tailored to each model, including one for the French made MF130; likewise trailer hydraulic pipe kits were available across the range.

A Belt Pulley and Guard kit was sold in two versions, a standard-duty version for the MF130 and 135 and a heavy-duty one for the higher horsepower models.

The Tractor Jack was still marketed in 1968, listed as suitable for the 135 and 165 – personally I think going up to that weight with that type of jack would be putting life and limb at risk!

Frowned upon today was the option, at extra cost, of the fitment of a cigarette lighter to all the Coventry-built 100 Series tractors.

A range of PAVT wheels was available as extras for those models not so equipped, and front and rear wheel weights were listed for most of the tractors in the range.

A Front Weight Frame could be had, incorporating a towing hitch and capable of taking up to eight 'Jerry Can' weights each of 60lb (27.2kg). This frame was suitable for all Banner Lane-built tractors, but for the MF130 and MF1100 different designs were needed.

Other smaller items were offered such as Dual Wheel Kits, rear view mirrors and brackets as well as an extension to the PTO control lever. Needless to say, accessories were offered for the implements but most are covered in the relevant part of Chapter 10.

An interesting fork-lift conversion of an MF135.

Chapter 12

The 100 Series Overseas

This chapter is based entirely on material provided to me by two people, John Farnworth and Bob Dickman, so let them unfold their personal recollections of MF100 Series tractors operating in overseas territories.

John Farnworth wrote:
Although I have not been a lifetime commercial farmer, I have farmed for very significant periods of my career. This includes at home, working on my father's and uncle's arable farms, working at the University College of North Wales' farm as a student, and then in my long overseas career in agricultural research and development I have established and managed four experimental farms in the Middle East. In most of these situations I operated the MF 100 Series tractor models. On reflection I have come to realise how the main period of my career in which I was involved practically with farming had in fact spanned the heyday of the 100 and 1000 tractor eras. It is a bit of a source of regret that I had no involvement with any of the 1000 series as they looked to be such dignified machines when I browsed the MF World Wide catalogues! These enormous tomes I first came across in Saudi Arabia where Ministry of Agriculture officials used to ask me what I thought would be appropriate for their harsh desert conditions.

The MF 100 Series always gave me great pleasure to operate. Somehow the full meaning and potential of the grey Ferguson tractors had been interpreted through into them but with of course increased horsepower models. I had always thought it a pity that Harry Ferguson appeared to have fought so shy of higher horsepower tractors for so long. The 100s were still virtually simple "step on" tractors like the old grey Ferguson.

Although my earliest childhood experiences on the family farms were of the old Wallis-style Massey-Harris tractors, my first solid driving experience was of my father's new TEA 20 Ferguson tractor, which was delivered in June 1948. This was a wonderful experience – not just driving a tractor, but exposure to the Ferguson System as well. A new concept in tractor design for a new generation – I was four when it arrived. Quite simply it all felt right, and as I grew up with this tractor so I became a devotee of the Ferguson System. We still have this tractor, serial number TEA39615, in the family.

Time moved on. At home the MF35s had their era and I eventually went off to university. On one undergraduate orgy we went on a trip to the 1964 Royal Smithfield Show in London, at which the 100 Series was launched in the UK. It certainly was spectacular. A load of glitzy razzamatazz for a group of absurdly well polished tractors. But how splendid they were. Massey Ferguson had at last come up with a new model range – and of striking new design.

The university farm where I was a student were soon to acquire an MF165 to replace their MF65. MF had some deal going through their agents whereby, to gain advertising, educational establishments could lease tractors on exceedingly favourable terms. This particular one had the early fibreglass-type cab. It all seemed such luxury. A large tractor, comfortable cab, the Ferguson System and plenty of power. I occasionally drove it when they were short-staffed, or at lunchtimes during hay or silage times. One really felt you were in charge of the job on one of these machines. And it could march up some of the steeply sloping fields with a 12-inch three furrow plough.

Another tractor that I drove on the university farm was an MF135. This was a contractor's tractor. He came each year with his 135 and an MF711 Potato Harvester. The contractor used to get fed up with driving and he would swap jobs with me and go on the harvester. The combination was a delight to drive. The tractor had Multi-Power which was useful for coping with variable loads of potatoes coming up to the picking table, and, using the tractor's hydraulics, digging depth was easily adjusted or the machine shifted slightly to left or right to accommodate varying drill width or offset of the drill.

Life moved on. My first job took me overseas to Saudi Arabia to expand and develop a major experimental farm for the Saudi Arabian government. It was too good to be true. On arriving I saw two brand-new 165s in the yard. I settled in, biding my time as the new manager and getting everything and everyone sized up before settling on action plans. The small farm I had inherited was ticking over in mostly traditional, manual-labour fashion with a clumsy old International B614 being used for a bit of haulage work. Something of a gutless tractor, I thought, and with an appallingly heavy clutch. There was also a yard full of new but abandoned MF implements in various states of being pirated for spares for other locations. Some had started to rust badly in the salt-laden soil. However, the potential for putting the Ferguson System of tractors and implements to good use here was clearly a star to be followed. But little did I know that all this equipment was registered in the Saudi government stores system. It all required

In Saudi Arabia, the 165 easily pulled this subsoiler to full depth with the help of water ballast tanks fitted to the rear wheels. This is Fahid, my best driver of all.

The MF20 with trailed hydraulically-operated land plane.

"releasing". By way of a brief explanation, the Saudi civil service rules were derived from what the British had applied in India. The rules had become an excuse for bureaucrats to do nothing – which is of course the safe way. Some four months passed and I had made no progress in getting these tractors released. My requests had gone up to ministerial level and back to local storemen several times. Finally I decided to defy the bureaucrats. I bought a couple of batteries and spare sets of keys from the MF agency in the capital Ryadh and away we went. The farm never looked back. We reclaimed many acres of salty desert land, achieved some world-record yields of year-round irrigated alfalfa and grass crops, and all with the Ferguson System applied through an increasing fleet of MF165s and a 135 (MF20) industrial.

It was in this desert environment that it was brought home to me just how good the engineering of these tractors was. To push the job on I used to do a lot of tractor driving in the afternoon (after official government working hours) for several years – land reclamation is a time-consuming process especially

Demonstrating a Malleti rotary trencher on an MF165 which had been fitted with a creeper gearbox.

Making a perfect seedbed with a Howard E4 rotavator which I had a local blacksmith make before Howard offered it as an accessory (about three months later!)

Abdullah Aloowi cut all the farm's forage crops 365 days a year with an MF51 Rotary Disc Mower. Note the wire mesh around the sunshade to protect his teeth from flying pebbles!

when all-year-round irrigation is involved as well. There was repeated heavy work involved in land levelling, drainage, sub soiling, chisel ploughing and rotovating. For six summer months we had maximum temperatures around 45°C. Fortunately the tractors were fitted with MF sunshades but the gearbox got extremely hot between your legs! I used to cover it in a hessian sack and drip water over it when I was driving. Imagine the load on those Perkins engines pulling a subsoiler – we had homemade wheel weights (water tanks) and ballast in the tyres. But I can never recall making a tractor boil. I was astonished by this, the more so when, browsing the tractor spare parts book, I discovered that an eight-blade fan was available for "tropical" work. We were in the arid subtropics which are renowned for being even hotter! Our tractors had reached Saudi Arabia fitted only with the standard four-blade fan and thrived on it. The tractors were truly amazing – they had no persistent or nagging types of breakdowns. Most repairs were fair wear and tear and fixed by our team of local Saudi mechanics, who I can proudly report were second to none for initiative and sensible improvisation. I can't recall a single instance of hydraulic system, engine or gearbox failure.

We were also short of good drivers. I was supposed to use "official" ministry drivers but they were beyond description. Lazy, opinionated and, worse, they refused to adopt the Ferguson System. They were rejects from the national oil company ARAMCO. They had an unbreakable habit because they had all been initiated on bulldozers shifting sand dunes for the oil company, but not being able to stand the pace had taken lower-paid Ministry of Agriculture employment and an easier life. The bulldozers of the day all required repetitive up-down hand operation

Collecting bales of freshly baled alfalfa. Saleh at the wheel of a 165 fitted with an MF sunshade.

A simple but valuable implement, an MF Border Ridger for forming the bunds between irrigated fields.

The Hofuf oasis had maybe thousands of acres of drainage area to which the saline drainage waters ran. These were invaded with Phragmites reeds. John's project was contemplating reclaiming some of this area but he needed soil samples first. Here is John Pritchard on the MF20 using an MF Post Hole Borer.

of the hydraulic lever to control blade depth at the front of the machine. So, given a Ferguson System of hydraulics which controlled matters automatically and at the back of the machine, they were lost in a big way. They could not break the bulldozer habit and the draft control lever was incessantly twitched by them. Worse, they arrived late for work, went home early, frequently slept on the job and had interminable tea breaks, which couldn't be taken in the field but had to be taken back in the yard. I became frustrated at the lack of progress so I decided to shatter tradition. I had it on good word from my wonderful field foreman Saad Eneenya that he could select labourers who could become drivers and would work for the same money – the theory being that they would be grateful to be rid of some of the toil of labouring. Miraculously I got the initiative past my truly excellent Saudi co-manager, who was the head of the farm's bureaucracy. We never looked back, despite a seven-year ministry internal legal wrangle festering on about the employment,

Ace driver Fahid operating a Howard Rotoseeder for seeding oats into semi-dormant tropical Rhodes grass to obtain more productivity over the cooler winter months.

Mixing concrete with an MF135 at 7500 feet up on the Yemen highlands inter-montane plains in Dhamar Province. My first job was to get a housing compound erected for the project staff.

My son Trystan inspecting the soil smoothing work on the non-irrigated dry land cropping area.

Helping out won the dairy farm next door, which belonged to North Yemen's President Al Saleh. Here, forming the bunds for border strip irrigation.

Using the 135's PTO to operate an Alvan Blanch Thresher.

My best driver Ali Saleh demonstrating his MF135 to local farmers at a project open day.

PTO drive again, to operate an Alvan Blanch Maize Sheller.

status and "legalities" of labourers driving. The "official" drivers continued to draw their inflated salaries for doing nothing, probably till retirement. But at least I and the farm thereafter progressed with ever -improving driver proficiency with the Ferguson System.

We progressed so rapidly that we were short of tractors for the increasing tasks on the farm and our budget couldn't acquire tractors fast enough. In came another initiative: we raided the machinery yards of other moribund experimental farms (there were many) and pulled together a selection of abandoned 35 and 65 tractors which I could see only needed a few spare parts, batteries or tyres to bring them back to life.

A modified MF24 chisel plough fitted with seeder tubes behind each leg for sowing sorghum on the dry land areas. The seated operator dribbled seed into a seed box which delivered seed down to behind the shares.

There were two strings of good fortune here. Firstly Saudi storemen actually seemed to like transferring equipment between different locations, secondly I had tapped into an apparently bottomless budget for spare parts. We used the 35 and 65s for generally lighter duties, but some 65s were committed to gruelling water pumping duties in summer – 24 hours a day on the PTO. Our station became renowned for the way we salvaged and kept equipment running. It was a wonderful advertisement for Massey Ferguson, which sadly was under exploited at the time by both the Saudi MF agent and MF themselves.

Breaking up freshly ploughed dry land with a Cousins Levelling Harrow. Driver Ali Saleh being assessed by foreman Ahmed Fahran.

Some new tractors we actually imported direct from the UK. They were trucked in by Astram, a transport company who pioneered the route from the UK to the Gulf region. It was quicker and much cheaper than buying them from the Saudi MF agent at the time. This agent subsequently did in fact try to not supply us with spare parts for them!

We bought in three MF 165s and an MF 20, the industrial version of the 135. We had no specific requirement for an Industrial but at that time 135s were in very short supply in the UK. Our fleet of 165s was a mix of MkIs and MkIIs with respectively the 203 and 212 engines. The 212s had a useful bit of extra power above that of the 203s. We also had a Ford 5000 given to us by the Ministry but like the International B614 I thought that it too lacked torque, which was to be found aplenty in the MFs.

On some particularly bad land we had to pull a single-leg subsoiler with two 165s – just a chain between the two – and we could get down to a metre depth in 3rd gear, low range. They would chug away relentlessly in the blistering heat. But I was being watched. A ministry-employed agronomy graduate who had been placed to work with us decided it was his day to make his mark in history and curry favour with his employers. He reported me, claiming that I would pull the rear tractor in half! I was summoned to high places in the capital Ryadh for an explanation. I pointed out that we had been doing this for a couple of months now and no harm had come to tractor or man. No one was convinced. Putting all my faith in MF engineering I gave a guarantee that if indeed the tractor broke in half then I would personally pay for the repairs. I heard no more of it and mercifully never had to pay out. My excellent chief mechanic Mohammed Abdulgader laughed off the stupidity of the agronomy graduate. Mohammed was wiser: he ensured that the engine oils were all changed at twice the recommended frequency, which was no problem in a country spilling over with oil.

I suppose that on reflection one of the great pleasures of operating the tractors in a desert environment was that most of the tractors were without cabs, though some had simple sunshades. It brought back the Ferguson connection – driving in the open air. This open-environment driving experience was to be repeated later in Yemen and Oman. We did have to modify the sun shades in Saudi Arabia on tractors operating the MF disc mowers to cut Alfalfa. This followed a driver having a front tooth knocked out by a flying pebble. We paid for a new gold tooth for him and wrapped the sides of the shade in fine wire mesh, which was readily available and commonly used for mosquito netting on house doors and windows.

On my first annual leave I went home to find that my father had bought his first 100 Series tractor, a 165, a rare occasion on which son was able to lecture father about matters pertaining to and, importantly, based on experience. He had that tractor until he retired and over the years he and my uncle were to add a 135, 178 and 185 to their tractor fleet as well as two of MF Industrials. These all stayed with them till their retirement in 1990. I didn't drive it much but I became particularly fond of the 178 with Multi-Power, PAVT wheels and cast iron wheel centres. There was a quiet and powerful lugging dignity about this tractor. On his retirement a chap who had worked for him bought my dad's 178. It had been "his" tractor when he worked for my dad on mainly baling and ploughing, and it is still with his sons today. I hope one day to be able to buy it back to add to our tractor collection.

However, in truth I never enjoyed driving the family tractors as much as those I had driven to date or was to experience later overseas. Why? Quite simply they had cabs – some rigid, some of the canopy and frame type, both types made by MF. Driving a tractor just isn't the same experience without the wind, sun, rain or whatever in one's face and driving at one's body, no matter what the extremes. I do however acknowledge that, in the quest for productivity from drivers, cooled or heated cabs are now a pre-requisite for farming's economic viability. But when I look back on my tractor driving days what do I recall with most affection? Well, it's the cab-free times such as ploughing in a blizzard at home on Barton Moss, the day it got up to 48 degrees C when I was subsoiling in Saudi Arabia, or early morning with air-frosted fingers in North Yemen in the crisp dawn light at high altitude.

I moved on from Saudi Arabia to a site 7500ft up on the inter-montane plains in North Yemen. The scale of agriculture was smaller here and not backed by lashings of oil money. They practised both irrigated and rain-fed agriculture, the latter being the major area. Here I had two 135 tractors which were fitted with 14in wide rear tyres and shell mudguards. No cabs and no sunshades. I viewed them as the ultimate development of the old grey Ferguson tractor. Ample power yet they were still simple step-on tractors. I really couldn't detect any power loss at this altitude which I thought was amazing. Our petrol-engined Land Rovers certainly suffered. My local drivers loved the 135s and indeed they were one of Yemen's favoured tractors in those days, with many bought for their family farms by labourers returning from working in the oil-rich Gulf. We bought several items of MF and other manufacturers' cultivation equipment to undertake all the tasks needed for the very diverse range of activities on an experimental farm. The Ministry there also issued us with a Kubota four-wheel-drive tractor. Whilst the extra traction was good for the odd bit of subsoiling, the lack of refinement and clutter of controls contrasted sharply with the simple refined efficiency of the 135s. The lugging power of the 135s' Perkins engines was quite superior to whatever was fitted in the Kubota – it seemed to be all revs and no guts.

We used to fit cage wheels for seeding the terraces in Yemen immediately after rains in order to prevent deep wheel marking. My young son loved to ride on the tractor with me (one of the pleasures of living in a country with little regard for safety!) in the afternoon when I was doing a bit of cultivation after the staff had gone home. Anyway, we were travelling quite fast with implement raised along our dirt road when up went son's arm and a shout of "Dad!" I was looking left at the time but he to the right. He just wanted to point out that the cage wheel had come off and was overtaking us!

In Yemen I was fortunate enough to have a superb self-trained mechanic, Mohammed Saleh. He was an older man who had been trained by the British army in Aden (which became South Yemen).

Ethiopia: The cotton seeding brigade crossing the River Awash on a Bailey Bridge. Originally a ferry was used.

The cotton seeding brigade lined up ready to start the sowing season with either MF Plate planters or Stanhay Seeders.

Rotovating in the end of season cottton crop.

From Yemen on round the corner to Oman. By this time in the mid-1980s the era of the 100 Series tractors was drawing rapidly to a close. We had a large desert farm to develop for His Majesty the Sultan. There was a good MF agency in the capital, Muscat, and sadly I had to move with the times: 100 Series tractors were no longer available. Our first purchase was a couple of MF290 four-wheel-drives. Although they performed well, that seat-of-the-pants driving experience of the old Massey-Harris, Ferguson and early MF two-wheel-drive tractors, which had given me such pleasure had gone – forever, I fear.

Good friend Malcolm Valentine had for several years been the Farm Mechanisation manager on a 30,000-acre cotton project, the Tendaho Plantations Share Co. at Tendaho in the Awash valley in Ethiopia. He operated a fleet of about 150 mainly MF185 tractors which he thought the world of, but this included a few MF178s, 175s and twelve MF1080s. For many years they were the only tractors used on the project. This tractor fleet was of such significance that it was regularly visited by MF officials and he had very good support from Ries Engineering, the MF main agents in the capital Addis Ababa. Their origins went back to Harry Ferguson days, when the Emperor had a range of Ferguson tractors and implements.

The project had land on both sides of the river Awash. Malcolm set up a rigorous service schedule and driver training programme. Dust and battery failure were two major problems. The dust problem was solved by changing the air filter, oil or cartridges, twice a day. Battery problems were overcome by fitting

Ploughing in readiness for the coming cropping season. Note the safety frame by this time.

An MF185 pulling a set of Schintzhy heavy duty disc harrows.

Lucas wind-up impulse starters as I had done with my four-cylinder MF35s in Saudi Arabia. As for the drivers, they always wanted to go too fast and had no appreciation whatsoever of the dangers of speed, and particularly going round corners at high speed! Some drivers killed themselves and a few later tractors were supplied fitted with safety frames.

I have also worked on consultancy assignments in the Sudan, where MF have a long history. There were many MF165s working there in the 1970s. It is somewhat ironic that recently Sudanese buyers (and Kenyans) have come into the UK market for good secondhand MF165s to ship back home. Unfortunately I don't seem to have any photos of them at work in the Sudan.

My father always said the MFs had good resale value and I think that after the Ferguson System he would have put that as his No.2 reason for buying

A 185 with safety frame and home-made sunshade pulling a 1-ton Vicon trailed Fertiliser Spreader.

A sad day in 1990. Dad and my uncle hold their dispersal sale and away goes part of my own personal history and a lot of MF and other tractors.

them. In Saudi Arabia, whenever we finally persuaded the storemen to dispose of old farm equipment, the same experience applied – the locals were very keen on the MF equipment. One of our shrewd Saudi graduate trainees did not miss this point and went on to establish an MF sub-agency in the eastern province of Saudi Arabia.

My Dad and uncle retired in 1990 and their dispersal sale was a sad day for me. My folks had farmed the farms for three generations and I had been brought up on them. The final line-up of all the tractors for sale was a sad experience but the sale went well and they were well pleased with the outcome.

In retirement, life has come round in a full circle. I bought a small farm 14 years ago and needed a workhorse. What else could it logically be but an MF165? I watched the advertisements for a few weeks and spotted what seemed to be a good one up in Cumbria. It was love at first sight – an excellent specimen with no rust or dents, and the agricultural engineer owner had just renewed the brakes. Additionally it had no cab though it had originally

My Dad's MF165 spent many of its summers on spraying duties fitted with rowcrop wheels. It is seen here filling up with water at my uncle's farm

My uncle's MF178 Multi-Power seen after it was sold off at his dispersal sale to ex-employee Andy Jenkins, who used to drive it on baling and ploughing duties. Andy was a very careful driver and looked after it well, but sadly he died too early in life. His sons now own the tractor.

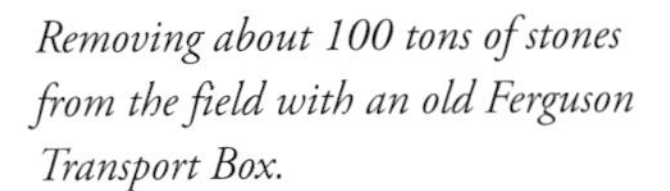
Removing about 100 tons of stones from the field with an old Ferguson Transport Box.

Ploughing the new field with a Ferguson Single-Furrow plough with disc removed was not a pretty sight!

been fitted with one. It is a Mk 1 with a Perkins 203 engine, of 1967 vintage, and was new to a farm in Wiltshire. Apart from oil changes, a new starter switch and a leaking front crankshaft seal I have not had to do anything with it and it is an excellent starter. In this last year I have used it to reclaim a field which we bought and which had never been cultivated in its history – more land reclamation! It comprised 50% eight-foot high gorse and 50% five-foot high bracken and brambles, very stony and with near-surface bedrock in many places. All were impossible to walk through. We had to clear the gorse with a JCB then slash all the bracken areas before we could do anything. This all left an enormous amount of trash on the surface. Most work was undertaken in third low gear but even so we broke several single-furrow plough points and cultivator legs – both Ferguson of course! I had to take the disc and skimmer off the plough because of all the trash. The resultant ploughing was not competition-winning but at least I had created soil! By hand I cleared probably in excess of 100 tons of stones from the field using a Ferguson transport box.

The first job of 2016 was to prepare a seedbed with a set of old Massey-Harris Trailing Disc Harrows.

Working in a four-ton per acre application of ground limestone with a Ferguson Tiller, marking the final job for 2015.

Delegates gathering outside the "temple" at the Massey Ferguson Olympiad in 1964.

Bob Dickman, who kindly contributed the photographs and the basis for the text in the next part of this chapter, first worked for Harry Ferguson and then for Massey Ferguson, starting his career as a trainee instructor at Stoneleigh, and then progressing to become fully qualified in that role. In these early days he was involved with students learning about the workings of the TE20 tractors and their implements, but within a short space of time he was moved on to evaluating a new training scheme to enable instructors to teach students the workings of the FE35 tractors that were being introduced to the farming community in the UK during October of 1956. Particular reference was given to the then new CAV DPA (distributor) type of fuel-injection pump. Attention was also focused on the two-lever hydraulic control system. Yet another area for special consideration was the functioning of the two-stage clutch fitted to the De Luxe models, another MF first, along with study of the six-speed gearbox, which had sliding spur gears as opposed to the constant-mesh helical-cut gears of the TE20 tractors.

After four and a half years in the role of training instructor Bob was moved to the Export Service Department at Fletchamstead Highway, Coventry, then under the direction of Charles Voss. In 1960, following another successful spell of training, he was given his first export assignment, working alongside MF's Sales Manager, Jaap Volgelsang, and they both attended the Leipzig Fair in February 1965 upon the introduction of the 100 Series tractors. Bob remained with The Export Service Department but in a more senior role, developing Distributor and Dealer After Sales Service Standards, which included premises, workshop layout, special service tools, signage, etc.

He states, "A great deal of resources and effort were given to markets apart from pure product and technical support". He adds the caveat, "At this time I first encountered company politics and the need to know the right people at the top end of the hierarchy in order to progress one's career."

Bob drew my attention to the MF World Wide launch of the 100 series, known as the Massey Ferguson Olympiad, in 1964, which he attended. It was held in the town of Lagonissi just outside Athens. Delegates and their wives representing all of MF Export Distributors from 110 countries attended the three-day event, which cost over £100,000 to put on, in co-operation with the host distributor J.D. Sarackis, who it was claimed was the sixth richest man in Greece. Bob kindly sent me six photographs of this event. Space precludes showing them all here but some of the captions he provided are well worth elaborating on. This event was the most ambitious and successful launch Bob ever attended in his 43 years with MF.

Bob outlined a major MF initiative introduced in 1980, firstly in Mozambique but over the years other countries in and out of Africa including Somalia, Ethiopia, Tunisia, Sri Lanka and Bangladesh. This initiative was named Refurbishment of Original Components, or ROC, which was a registered

The Greek play, starring the new MF135!

MF trademark, chosen because like that legendary bird in Sinbad the Sailor, it was very strong and lived to a great age. The programme came about through the realisation that there were thousands of battered MF100 Series tractors in "graveyards" in these developing countries. MF set up the ROC programme to enable local MF dealerships in these territories to bring these abandoned tractors back into working order, but to precisely set standards and incorporating the latest design modifications. These standards were "policed" by UK personnel, including David Stevens.

The programme gave the opportunity not only to upgrade the specification of the tractors but to enhance the level of training and expertise of the mechanics on the job, and also to promote the use of special service tools and simple test equipment for hydraulics and fuel-injection equipment.

Each tractor was given a new Perkins-built short engine and a new and upgraded gearbox, while new and upgraded hydraulic pumps were always specified. All wearing parts such as bearings, oil seals, steering joints, brakes and electrical equipment were renewed. Steering boxes and rear axles were fully reconditioned. Finally the tractors were repainted, new decals applied and complete new sets of tyres fitted. When completed to MF's satisfaction all carried a full warranty and new registration plates.

This was true recycling and it was estimated that a MF tractor thus refurbished was 25-30% cheaper than the cost of a new unit. It is impossible to establish the numbers of MF100 Series tractors that contributed to the success of the ROC scheme but the general consensus is something in the order of 10,000.

Bob completed his career with MF as Export General Service Manager.

Bob supplied these pictures of MF100 Series tractors at work around the soda lake, Lake Magadi, which is about 20 miles long and 2 miles wide and situated on the floor of the Rift Valley in Kenya at just under 2000 feet above sea level. Throughout most of the year it experiences tropical sun which gives a shade temperature of well over 100°F. The lake's existence is assured by springs which continuously replenish it with fresh supplies of alkaline liquor, known to scientists as sodium sesquicarbonate or "trona". This location is one of a few anywhere in the world, another being in Wyoming, North America.

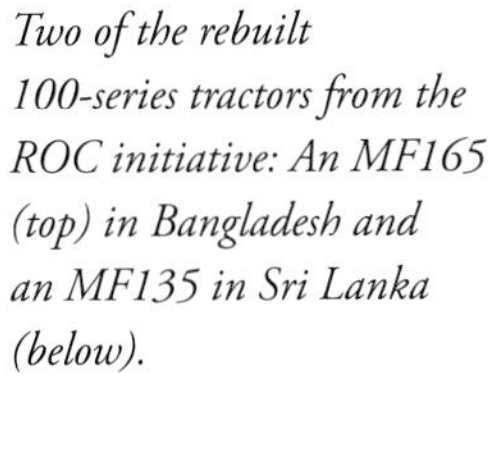

Two of the rebuilt 100-series tractors from the ROC initiative: An MF165 (top) in Bangladesh and an MF135 in Sri Lanka (below).

Salt water from this lake is pumped into concentration ponds by an axial flow pump driven by an MF135. By this means further evaporation of water takes place, raising the concentration to 13-14%. It is then transferred by another pumping operation into "making ponds". Here the solution lies at a depth of about 3 feet and with the intense heat most of the water is soon lost through evaporation. With alternated flooding and evaporation, several layers of salt build up, and when sufficient deposits have accumulated the salt is drawn manually into winrows and then loaded into 5-ton trailers drawn by MF135s to be taken to the processing plant.

On this site, in 1969, six MF135s were used for pumping and transport duties, while a further three MF165s were enrolled for access road maintenance.

Another of Bob's photos showing an MF101 self-

These poor quality photographs are included to give the reader an idea of the harshness of the Lake Magadi environment in Kenya.

The *MF101 Self Propelled Sugar Cane Harvester.*

propelled Sugar Cane Harvester, which was launched in Australia in time for the 1968 harvest; as can be seen (left) it is based on a reverse-mounted MF135. These machines were developed and manufactured at Crichton Industries Pty facility in Bundaberg, Queensland, which is near to Australia's sugar cane belt. The firm became a subsidiary of MF in 1966.

This Brighton Beach Cleaner was a British design and prototype tested in the UK using an MF135. Following successful development here it was tested and made in Australia and found to be capable of cleaning an acre per hour. It was noted in the Australian trials at Bondi Beach, Sydney, that it worked well picking up beer cans and bottles, but other beach trash such as condoms tended to block up the grill and other rotating components.

An MF2135 industrial tractor, equipped with half tracks at the rear and a ski option to replace the front wheels, was prepared by MF Queens Bridge Motors in Victoria, Australia, and shipped to the Australian Antarctic base at Mawson.

The MF Brighton Beach Cleaner.

An MF2135 about to be loaded aboard the Mella Dan *bound for the Antarctic.*

Chapter 13

MF100 Series Tractors in Use Today

Retired dairy farmer Robert Perry and his semi-retired MF135 deal with some logging duties.

As we all know, many MF35s and to a lesser extent MF65s are in use with a wide range of people needing a small and compact light tractor, but their successors in the 100 Series are even more sought after.

Although the 100 Series offers a much wider range of models in terms of size and power outputs, the MF135 is by far the most popular choice. So who are the users and buyers of these 40-50-year-old classics? Well, there are collectors aiming to have a representative of each model and variant within the range, thus highlighting the differences between them. There are enthusiasts who enjoy taking part in road runs and rallies. Then there are those who use their MF100 Series tractors for real work, perhaps a dairy farmer using it for slurry scraping duties or for light field work. Equestrian centres tend to use them too.

An MF135 is also, for example, an ideal machine for those who cut and supply logs for wood stoves. They might cut cordwood from hedgerow or woodland and cart it to their yard by trailer or link

An MF135 and a Massey Harris binder harvesting wheat reed to be used for thatching.

box, then attach a Ferguson cordwood saw to the three-point linkage or use a linkage-mounted PTO-driven saw bench to slice the cordwood to suitable log size, depositing the logs back into the trailer or link box for delivery to the customer. Perhaps they have invested in a tractor-mounted hydraulic log splitter to deal with the larger pieces.

Yet another type of work where these smaller 100 Series tractors are put to seasonal use is in the maintenance of local sports fields, which in recent years seems to have fallen more and more to volunteers with a community spirit.

The great benefit of using these classic tractors in whatever role is that they are far less costly to buy than a new compact tractor of comparable size. Most spare parts are freely available and at reasonable prices. Likewise they can be maintained and repaired by any competent DIY mechanic.

It is interesting to reflect that today the Indian arm of AGCO Tractor and Farm Equipment, TAFE, still manufactures some models that are very close in specification to the MF135, proving beyond doubt that these smaller tractors have an important role to play, particularly in developing countries of the world, a world where food production is of paramount importance.

Chris Massingham's beautifully restored MF20 ready for action with a topper.

The Indian-built TAFE tractors are very close in specification to the MF135.

Phil Kirk tackling the really long stuff with his MF135 and topper.

The 100 series tractors are very popular on tractor runs. Here we have an MF165 (top right), an MF188 (left) and a MF135 (above).

Appendix

100 Series Production figures

MF135

Year	Serial numbers	Units
1965	101 - 30282	30282
1966	30283 - 67596	37314
1967	67597 - 93304	25708
1968	93305 - 117428	24124
1969	117429 - 141425	23997
1970	141426 - 162200 End	20775
	Sub Total	**162,200**

Updated model

Year	Serial numbers	Units
1971	400001 - 403517	3517
1972	403518 - 419582	16064
1973	419583 - 432708	13126
1974	432709 - 445601	12893
1975	445602 - 457865	12264
1976	457866 - 469334	11469
1977	469335 - 479191	9857
1978	479192 - 487349	8158
1979	487350 - 490714	3365
	Sub Total	**90,713**
	TOTAL MF135	**252,913**

MF148

Year	Serial numbers	Units
1972	600001 - 602152	2152
1973	602153 - 604448	2296
1974	604449 - 605577	1129
1975	605578 - 607700	2123
1976	607701 - 609158	1458
1977	609159 - 609968	810
1978	609969 - 610892	924
1979	610893 - 610982	90
	TOTAL MF148	**10,982**

MF165

Year	Serial numbers	Units
1965	500001 - 512206	12206
1966	512207 - 530824	18618
1967	530825 - 547383	16559
1968	547384 - 563700	16317
1969	563701 - 581456	17756
1970	581457 - 597745	16289
	Sub Total	**97,745**

Updated model

Year	Serial numbers	Units
1971	100001 - 103621	3621
1972	103622 - 116352	12731
1973	116353 - 126447	10095
1974	126448 - 135035	8588
1975	135036 - 145431	10396
1976	145432 - 155686	10255
1977	155687 - 164416	8730
1978	164417 - 173143	8727
1979	173144 - 173696	553
	Sub Total	**73,696**
	TOTAL MF165	**171,441**

MF168

Year	Serial numbers	Units
1971	250001 - 250004	4
1972	250005 - 252120	2116
1973	252121 - 245306	2186
1974	245307 - 255966	1660
1975	255967 - 258063	2097
1976	258064 - 259958	1895
1977	259959 - 260616	658
1978	260617 - 261102	486
1979	261103 - 261173 End	71
	TOTAL MF168	**11,173**

MF175 & 178

Year	Serial numbers	Units
1965	700001 - 705651	5651
1966	705652 - 714165	8514
1967	714166 - 722678	8513
1968	722679 - 732157	9479
1969	732158 - 740300	8143
1970	740301 - 747282	6982
1971	747283 - 753108 End	5825
	TOTAL MF175 & 178	**53,107**

Note: MF178 launched at the Royal Show at Stoneleigh in July 1967, production would have commenced at about serial number 720200.

MF175S (Export Only)

Year	Serial numbers	Units
1968	650000 - 652060	2060
1969	652061 - 653720	1660
1970	653721 - 656010	2290
1971	656011 - 657362 End	1352
	TOTAL MF175S	**7362**

MF185

Year	Serial numbers	Units
1971	300001 - 302832	2832
1972	302833 - 310397	7565
1973	310398 - 315218	4821
1974	315219 - 319922	4704
1975	319923 - 326108	6186
1976	326109 - 332106	5998
1977	332107 - 335210	3104
1978	335211 - 339754	4544
1979	339755 - 340096	342
	TOTAL MF185	**40,096**

MF188

Year	Serial numbers	Units
1971	350001 - 350005	5
1972	350006 - 353295	3290
1973	353296 - 357062	3767
1974	357063 - 360782	3720
1975	360783 - 365086	4304
1976	365087 - 368349	3263
1977	368350 - 370155	1806
1978	370156 - 371305	1150
1979	371306 - 371333	28
	TOTAL MF188	**21,333**

The last MF100 Series tractor of the Coventry line – it looks to be an MF185 with inner and outer front wheel weights, rears cast PAVT centres, shod with what appears to be 15-30 tyres. The tractor is surrounded by shop floor workers and some white coated staff who had probably sneaked off from the drawing office.

Acknowledgements

There are a lot of people whom I have to thank for their help with this book.

John Farnworth's book *A Worldwide Guide to Massey Ferguson 100 and 1000 Tractors* is a unique reference for basic specifications of all models produced. It was he who helped me enormously to construct the text for Chapter 1. John also made a major contribution to Chapter 12, relating his experiences of operating MF100 Series tractors in far-flung lands.

David Walker worked for MF for 21 years, initially as a technical author writing instruction books and workshop manuals, then later as a service engineer, before moving on to become a regional Service Manager in both the UK and Europe. At one point he also dealt almost exclusively with the early MF articulated tractors. He diligently read my drafts and corrected my errors as well as making many helpful suggestions and additions, greatly improving the accuracy of this book: any errors are mine!

Bob Dickman contributed the photographs and related notes that I have used as part of Chapter 12; these represent just the tip of the iceberg of Bob's worldwide experience in a senior role for MF, a time span of 43 years.

Hans-Gören Persson from Sweden kindly sent me photographs of MF100 series forestry tractors and the aircraft tug. Peter Smith and Julia Browning provided pictures of some of their American 100 Series tractors. Photographs were also gratefully received from Thomas Neeb of his MF133 and Hubert Heiss of his MF177.

Jeremy Burgess, who wrote the Foreword to my book *Massey Ferguson 35 & 65 Models in Detail*, has helped me enormously with this project. His lengthy involvement with MF enabled him to put me in contact with Harold Lang in the USA, who worked for MF for nearly 38 years and kindly supplied some photographs for Chapter 1.

George French, a long-term MF enthusiast and professional, made a good many necessary comments on my drafts, as well as allowing us to take photographs of his MF148.

Jonathan Lewis made available three of his MF tractors – an early 135, a 168 and a 188 – for photography and helped us with captions for them.

Bernard Curtis was kind enough to let us photograph his well-used MF168 4WD conversion.

Robert Perry brought his MF135 and saw bench along for photography (and in the process provided me with some free logs!).

Thanks must also go to Nick Hill for the picture of his MF135 and Massey Harris binder in chapter 13.

I am very grateful for Ted Everett's keen interest in this project. Ted, who in his youth worked for Harry Ferguson as a photographer, moved on into the Massey Harris era as Chief Photographer and Archivist. Now well past retirement age, he still puts in three mornings a week at Stoneleigh ensuring that requests from writers, TV companies and editors of tractor magazines are met. It was he who kindly sorted out lots of archive pictures that I have used in this book.

My thanks must also go to Gail McKechnie for taking my handwritten drafts chapter by chapter, tweaking the poor spelling and putting the chapters into a digital format, as she did for my last book.

I am very glad that Andrew Morland consented to carry out the commissioned photography. His work makes a most positive contribution to this book.

Finally, thanks to Alison Harding for support and encouragement throughout this project.

Mike Thorne
Coldridge
July 2017